Praise for
THE UNIVERSE IN YOUR HAND

A *Shelf Awareness* Best Book of 2016

"*The Universe in Your Hand* is a masterpiece of popular science writing." —*Shelf Awareness* (starred review)

"Deft and dazzling." —*Publishers Weekly* (starred review)

"The reader will come away from [*The Universe in Your Hand*] with a deeper understanding of how modern physics has brought us closer to an ultimate understanding of reality." —*The New York Times Book Review*

"An epic feast of modern physics." —*The Wall Street Journal*

"Galfard dispenses with mathematical formulas in this foray into modern physics, making a lively imagination the only portal necessary for general readers hungry for the intellectual excitement of astral and atomic exploration." —*Booklist* (starred review)

"If Ms. Frizzle were a physics student of Stephen Hawking, she might have written *The Universe in Your Hand*, a wild tour through the reaches of time and space, from the interior of a proton to the Big Bang to the rough suburbs of a black hole. It's friendly, excitable, erudite, and cosmic." —Jordan Ellenberg, *New York Times* bestselling author of *How Not to Be Wrong*

"How can we ever hope to fully grasp the infinite complexities of space and time, or even the very nature of reality itself? Thankfully, Christophe Galfard shows us the way, through a well-written and thoroughly approachable journey through what we know of modern astronomy and physics. Part personal conversation, part travelogue, and part science primer for the non-scientist, *The Universe in Your Hand* is a delightful and highly educational read."

—Jim Bell, author of *The Interstellar Age* and *Postcards from Mars*

Christophe Galfard

The Universe in Your Hand

A Journey Through Space, Time, and Beyond

FLATIRON
BOOKS
NEW YORK

www.flatironbooks.com

The Library of Congress has cataloged the hardcover edition as follows:

Names: Galfard, Christophe.
Title: The universe in your hand : a journey through space, time, and beyond / Christophe Galfard.
Description: New York : Flatiron Books, 2016. | Includes index.
Identifiers: LCCN 2016005257 | ISBN 9781250069528 (hardcover) | ISBN 9781250069535 (e-book)
Subjects: LCSH: Cosmology—Juvenile literature. | Universe—Juvenile literature.
Classification: LCC QB983 .G35 2016 | DDC 523.1—dc23
LC record available at http://lccn.loc.gov/2016005257

ISBN 978-1-250-07641-0 (trade paperback)

Our books may be purchased in bulk for promotional, educational, or business use. Please contact your local bookseller or the Macmillan Corporate and Premium Sales Department at 1-800-221-7945, extension 5442, or by e-mail at MacmillanSpecialMarkets@macmillan.com.

Originally published in Great Britain by Macmillan, an imprint of Pan Macmillan.

First Flatiron Books Paperback Edition: April 2017

10 9 8 7 6 5 4 3 2 1

To Marius & Honoré

Contents

Foreword 1

Part One The Cosmos 3

Part Two Making Sense of Outer Space 51

Part Three Fast 115

Part Four A Dive into the Quantum World 147

Part Five To the Origin of Space and Time 201

Part Six Unexpected Mysteries 257

Part Seven A Step Beyond What Is Known 327

Epilogue 359

Acknowledgments 369

Sources 371

Index 373

Foreword

Before we start, there are two things I would like to share with you.

The first is a promise, the second is an ambition.

The promise is that the book contains only one equation.

Here it is:

$$E = mc^2$$

The ambition, my ambition, is that in this book I will not leave any readers behind.

You are about to start a journey through the universe as it is understood by science today. It is my deepest belief that we can all understand this stuff.

And that journey begins a very long way from home, on the other side of the world.

Part One

The Cosmos

1 | *A Silent Boom*

Picture yourself on a faraway volcanic island on a warm, cloudless summer night. The surrounding ocean is as still as a lake. Only the tiniest of waves wash against the white sand. All is quiet. You are lying on the beach. Your eyes are closed. The warm, sun-baked sand heats up air saturated with sweet, exotic scents. There is peace all around.

A wild shriek in the distance makes you jump and stare into the darkness.

Then: nothing.

Whatever shrieked is now quiet. There is nothing to be afraid of after all. This island may be dangerous for some creatures, but not for you. You are a human, the mightiest of predators. Your friends will soon be joining you for a drink and you are on holiday, so you lie back on the sand to focus on thoughts worthy of your species.

A myriad of tiny lights flickers throughout the vast night sky. Stars. Even with the naked eye you see them everywhere. And you remember questions you had as a child: what are they, these stars? Why do they flicker? How far away are they? And now you wonder: will we ever *really* know? With a sigh, you relax back on the warm sand and put these silly questions aside, thinking, why should we care?

A tiny shooting star gently streaks across the sky overhead and, just as you are about to make a wish, the most extraordinary thing happens: as if to answer your last question, 5 billion years suddenly pass and the next thing you know, you are no longer on a beach, but in outer space, floating through emptiness. You can see and hear and feel, but your body is gone. You are ethereal. Pure mind. And you don't even have the time to wonder what just happened or to shout and call for help, for you are in the most peculiar of situations.

In front of you, a few hundred thousand miles ahead, a ball is flying against a background of tiny distant stars. It glows with a dark orange light, moves towards you, spins. It doesn't take you long to figure out that its surface is covered with molten rocks and that what you are facing is a planet. A liquefied planet.

Shocked, a question comes to your mind: what monstrous source of heat could liquefy an entire *world* like this?

But then a star, immense, appears to your right. Its sheer size, compared to that of the planet, is just astounding. And it spins too. And it also moves through space. And it seems to be growing.

The planet, although much closer, now looks like a child's tiny orange marble facing a gigantic ball that continues to grow at an astonishing rate. It is already twice the size it was a minute ago. Presently, it has a red hue, and it angrily ejects huge filaments of million-degree-hot plasma that blast through space at what seems to be very close to the speed of light.

Everything you see is of a monstrous beauty. In fact, you are living through one of the most violent events the universe can provide. And yet there is no sound. All is silence, for sound does not spread in the vacuum of space.

Surely the star won't be able to keep growing at this rate; and yet it does. It is now beyond any size you could have imagined and the liquefied planet, pounded by energies beyond its strength, is blown to nothingness. The star did not even notice. It keeps growing, reaches about a hundred times its initial size and then, quite suddenly, it explodes, firing all the matter it was made of into outer space.

A shock wave passes through your ghostly form, and then only dust remains, blown in all directions. The star is no more. It has become a spectacular and colourful cloud that now spreads into the interstellar void at a velocity worthy of gods.

Slowly, very slowly, you come back to your senses and, as you realize what just happened, a strange lucidity fills your mind with a fearsome truth. The star that died was not a random star. It was the Sun. Our Sun. And the molten planet that vanished within its brightness was the Earth.

Our planet. Your home. Gone.

What you witnessed was the end of our world. Not a speculative end, not a far-fetched fantasy of supposedly Mayan origin. The real one. One that mankind has known would happen since some time before you were born, 5 billion years before what you just saw.

As you try to pull these thoughts together, your mind is instantly sent back to the present, inside your body, on the beach again.

Your heart races and you sit up and look around, as if waking from a strange dream. The trees, the sand, the sea and the wind are there. Your friends are on their way. You can see them in the distance. What happened? Did you fall asleep? Did you dream what you saw? An uncanny feeling spreads throughout your body as your queries start to shift: could it have been real? Will the Sun really explode one day? And if so, what will happen to humanity? Can anyone survive such an apocalypse? Will everything up to the very memory of our own existence vanish into cosmic oblivion?

Gazing once more up at the starlit skyscape above, you desperately try to make sense of what happened. Deep down, you know that you did not just dream it all. Although your mind is back on your beach, reunited with your body, you know you really *did* travel beyond your time, into a faraway future, to see something no one should ever see.

Slowly breathing in and out to calm down, you start hearing strange noises, as if the wind, the waves, the birds and the stars are all whispering a song that only you can hear, and you suddenly understand what they are all singing about. It is both a warning and an invitation. Of all possible futures available, they murmur, only one path will allow humanity to survive the inevitable death of the Sun and most other catastrophes.

The path of knowledge, of science.

A journey open to humans only.

A journey that you are about to take.

Another wild shriek pierces the night, but you hardly hear it this time. As if a seed just planted in your mind was already starting to sprout, you feel the urge to find out what is known about this universe of yours.

Humbly lifting your sight again, you now gaze at the stars with the eyes of a child.

What is the universe made of? What lies in the vicinity of the Earth? And beyond? How far can one look? Is anything known about the universe's history? Does it even have one?

As the waves gently wash over the shore, as you wonder if one will ever be able to probe these cosmic mysteries, the twinkling of the stars seems to lull your body into a half-conscious state. You can hear your approaching friends' conversations but, strangely, you already feel the world differently than you did a few minutes ago. Everything seems somehow richer, more profound, as if your mind and body were both part of something much, much bigger than anything you had ever thought of before. Your hands, your legs, your skin . . . Matter . . . Time . . . Space . . . Intertwined fields of forces all around you . . .

A veil you didn't even know was there has been lifted from the world to reveal a mysterious and unexpected reality. Your mind yearns to be back among the stars, and you have the feeling that some extraordinary journey is about to take you very far away from your home world.

2 | *The Moon*

If you're reading this, it means you've already travelled 5 billion years into the future. A good start, by anyone's standards. So you should be confident that your imagination is working well, and that is perfect, because imagination is all you'll need to travel through space and time and matter and energy, to discover what is known about our reality from an early twenty-first-century perspective.

You didn't ask for it, but you did happen to see what fate awaits mankind, indeed all life forms on Earth, if nothing is done to understand how nature works. To survive in the long run, to avoid being swallowed by a furious dying Sun, our only chance is to learn how to take our future into our own hands. And for this to happen, we need to unravel the laws of nature itself, and learn how to put them to good use. It's fair to say that we have a lot to get through. In the following pages, however, you shall see pretty much everything that is known so far.

Travelling throughout our universe, you will discover what gravity is about and how atoms and particles interact with each other without ever touching each other. You will find out that our universe is mostly made up of mysteries and that these mysteries have led to the introduction of new types of matter and energy.

And then, once you've seen everything that is known, you will jump into the unknown and see what some of the brightest theoretical physicists of today are working on to explain the very strange realities we happen to be a part of. Parallel universes, multiverses and extra dimensions will enter the picture. After that, your eyes will probably be shining with the light of knowledge and wisdom that mankind has been gathering and improving for millennia. But you should prepare yourself. Discoveries made during the past decades have changed everything about what we believed to be true: our universe is not only unfathomably bigger than expected, it is also immensely more beautiful than any of our ancestors could have imagined. And while we're at it, here is another piece of good news: to have figured out as much as we already have makes us humans different from all the other life forms that have ever lived on Earth. And that is a good thing, for most of the other life forms became extinct. The dinosaurs ruled the surface of our planet for about 200 million years, whereas we have done so for no more than a few hundred thousand. They had plenty of time to start questioning their environment and figure a few things out. They didn't. And they died. Today we humans could at least hope to detect a threatening asteroid early enough to try to deflect it. So we already have some powers they did not have. It might be unfair to say it, but with hindsight we might thus link the dinosaurs' extinction to their lack of awareness of theoretical physics.

For now, you still are on the beach, though, and the memory of the dying Sun is still vivid in your mind. You don't have that much insight yet and, to be honest, the twinkling dots that stud the night seem utterly oblivious to your existence. The life and death of earthly species makes no difference to them whatsoever. It appears that time, in outer space, works on scales that your body cannot grasp. The entire existence of a species here on Earth probably lasts no more than a snap of the fingers for such distant shining gods . . .

Three hundred years ago, one of the most famous and brilliant scientists of all time, British physicist and mathematician Isaac Newton, the man who gave us gravity from Cambridge University, England, actually thought in such terms about time: for him, there was the time of humans, felt by us all and measured by our clocks, and there was the time of God, which is instantaneous, which doesn't flow. From the point of view of Newton's God, the infinite line of human time, stretching backward and forward into infinity, is but an instant. He sees it all in one blink.

You are not God, though, and as you watch the stars, as a friend of yours silently pours you a drink, the immensity of the task at hand starts to feel overwhelming. All this is too big, too far away, too strange . . . Where to begin? You are not a theoretical physicist . . . but you are not the type to give up either. You have eyes and your mind is curious, so you lie down on the sand and start by focusing on what you can see.

The sky is mostly dark.

And there are stars.

And in between the stars, your naked eye perceives a dim band that glows with a faint whitish light.

Whatever this light is, you know the band is called the Milky Way. Its width looks to be about ten times that of a full Moon. You stared at it many times when you were younger, but not that much recently. As you now see it, you realize that it is so conspicuous that it must have been known to your ancestors since forever, and you are right. Ironic to think that, after centuries during which men and women debated its nature, we now know what it is – although light pollution makes it invisible from most inhabited places.

From your tropical island, however, its presence is overwhelming and, as the Earth spins as the night advances, the Milky Way moves through the sky, like the Sun during the day, from east to west.

The possibility that the future of humanity lies somewhere out there, beyond the Earth's sky, starts to become real in your mind, and spellbinding. Focusing, you wonder if it is possible to see all there is in the universe with the naked eye. And then you shake your head. You know that the Sun, the Moon, some planets like

Venus, Mars or Jupiter, some hundreds of stars* and that fuzzy streak of whitish dust called the Milky Way don't add up to being Everything. There are mysteries hiding up there, out of sight, between the stars, mysteries that are just waiting to be unravelled . . . If only you could probe it all, what would you do? You would start with the vicinity of the Earth, of course, and then . . . then you'd shoot away and go as far as possible, and then . . . Your mind obliges!

As amazing as it sounds, your mind *does* start to move away from your body, upwards, towards the stars.

A spinning sensation of vertigo hits you as your body, and the island it is lying on, recede rapidly beneath you. Your mind, shaped as an ethereal you, is heading up, and east. How that is even possible, you do not have a clue, but there you are, higher than the tallest of mountains. A very red Moon appears, suspended above a distant horizon, and in far less time than it takes to say it, you find yourself out of the Earth's atmosphere, flying across the 236,000 miles of emptiness that separate our planet from our only natural satellite. From space, the Moon appears as white as the Sun.

Your journey through knowledge has begun.

As only a dozen humans have done before, you've reached the Moon. Your ghostly body is now walking on it. The Earth has disappeared below the lunar horizon. You are on its so-called *dark side*, the side that never sees our planet. There is no blue sky, nor any wind, and not only do you see many more stars above your head than from anywhere on Earth, they don't twinkle. All this because there is no atmosphere on the Moon. On lunar soil, space begins a millimetre above the ground. No weather ever erases the scars that

---

* It might feel like you can see millions of stars on a dark night; in fact, the human eye can make out just a couple of hundred from within a city, and between 4,000 and 6,000 in the countryside, away from light pollution.

scatter its surface. Craters are everywhere, frozen memories of what once hit that barren soil.

As you start walking towards the Earth-facing side of the Moon, the history of its birth magically pours into your eager mind and you stare, dumbfounded, at the ground beneath your feet.

What violence!

About 4 billion years ago, our young planet got hit by another one, the size of Mars, which tore a huge chunk of it off into space. During the following millennia, all the debris from that collision settled into a single ball in orbit around our world. When that was done, the Moon you are now standing upon was born.

Were it to happen today, such a collision would be more than enough to wipe out all life forms on Earth. At the time, though, our world was bare, and it is funny to think that without such a catastrophic bang, we would have no Moon to illuminate our nights, no significant tides, and life as we know it probably wouldn't exist on our planet. As the blue Earth appears in front of you, above the lunar horizon, you realize that catastrophic events, on a cosmic scale, can be for the best just as they can be for the worst.

Your home planet, seen from out here, is the size of four full Moons put together. A blue pearl floating in front of a black, star-sprinkled background.

Our world's true extent in space is, and will always be, a humbling sight.

And as you walk some more and watch it rise in the lunar sky, even though all seems quiet and safe, you already know better than to trust such an apparent peace. Time has another meaning out here and given the eons that continue to unfold, the universe's violence seems unavoidable. The craters that scar the Moon's surface are but a reminder of it. Hundreds of thousands of mountain-sized boulders, adrift in space, must have pounded it over the ages. And they must also have hit the Earth – but our planet's wounds have healed, for our world is alive and hides its past away deep under its ever-changing soil.

Still, within such a universe, you suddenly sense that your home world, despite its ability to heal, is fragile, almost defenceless . . .

Almost.

But not quite. It now has us. It has you.

Collisions such as the one that led to the birth of the Moon mostly belong to the past. Today there are no stray planets menacing our world, just loose asteroids and comets – and the Moon partly protects and shields us from such hazards. Danger, however, looms everywhere and, as you watch the Earth's blue-hued image hang in the dark sky, an extraordinarily bright ball of light suddenly rises behind you.

You turn around to face a star, the brightest and most violent object that can be found near our home planet.

We have named it the Sun.

It lies 93 million miles away from our world.

It is the source of all our power.

And as your mind becomes passionately overwhelmed by the sheer amount of light that emanates from this extraordinary cosmic lamp, you leave the Moon behind and start flying towards it, our local star, the Sun, to find out why it shines.

3 | *The Sun*

If mankind could, one way or another, harvest all the energy the Sun radiates in one second, it would be enough to sustain the entire world's energy-needs for about half a billion years.

As you fly closer and closer to that star of ours, however, you realize that the Sun is not as big as when you saw it 5 billion years in the future, as it reached its end. Still, it is big. To put things into perspective, if the Sun were the volume of a large watermelon, the tiny Earth would lie some 140 feet away – and you'd need a magnifying glass to see it.

You've reached a few thousand miles above the Sun's surface. Behind you, the Earth is but a bright dot. In front of you, the Sun fills half your sky. Bubbles of plasma are erupting all around. Billions of tonnes of beyond-hot matter are ejected right before your eyes and shoot through your ethereal body, as huge, seemingly random loops open up in the Sun's magnetic field. The scenery is extraordinary, to say the least, and, exhilarated by its power, you suddenly wonder what it is that the Earth lacks that makes the Sun so special. What makes a star a star? Where does its energy come from? And why on Earth does it have to one day die?

To figure this out, you head for the harshest place that can be, the heart of the Sun, more than 310,000 miles below its surface. The Earth, by comparison, is about 4,000 miles from surface to core.

As you jump head first into the bright furnace, you remember that all the matter we breathe or see or touch or feel or detect, even the matter your real body contains, is made of atoms. Atoms are the building blocks of everything. They are the Lego bricks of your environment, if you will. Unlike Lego, however, atoms are not rectangular. They are mostly round and consist of a dense, ball-shaped nucleus with tiny, distant electrons swirling around it. Like Lego, however, it is possible to classify atoms by size. The tiniest of them all has been called *hydrogen*. The second smallest has been named *helium*. Take these two atoms together, and you have about 98 per cent of all the known matter in the known universe. A lot, certainly, but a smaller proportion than it was in the past. Some 13.8 billion years ago, it is believed these two atoms accounted for near enough 100 per cent of all known matter. Nitrogen, carbon, oxygen and silver are examples of atoms that can be found today which are *not* hydrogen nor helium. So they must have appeared later. How? You are on your way to find out.

You dive deeper and deeper inside the Sun; the temperature rises and becomes mind-bogglingly hot. As you reach its core, we are talking 29 million degrees Fahrenheit. Maybe even more. And there are plenty of hydrogen atoms everywhere, although they have been stripped naked by the surrounding energy: their electrons are loose, leaving bare nuclei. The pressure is so high, these nuclei are so tightly packed by the weight the whole star exerts on its own heart, that they barely have any freedom to move at all. Instead, they are forced to fuse into one another to become a bigger nucleus. You see it happening right in front of you: a *thermonuclear fusion reaction*, the creation of big atomic cores out of smaller ones.

Once built, as they move out of the furnace that gave them birth, these heavy cores will team up with the lone, free-moving electrons that were stripped away from the hydrogen nuclei, and become new, heavier atoms: nitrogen, carbon, oxygen, silver . . .

For a thermonuclear fusion reaction to occur (the creation of big atoms out of small ones), a stupendous amount of energy is needed, and that energy is here provided by the Sun's crushing gravity, which effectively pulls everything towards its core, thus compressing it immensely. Such a reaction cannot take place naturally on (or inside) the Earth. Our planet is too small and not dense enough, so its gravity cannot make its core reach the temperature and pressure needed to trigger one. By definition, that is the main difference between a planet and a star. Both are roughly round cosmic objects but planets are basically small, with rocky cores that are sometimes surrounded by gas. Stars, on the other hand, can be viewed as huge thermonuclear-fusion power-plants. Their gravitational energy is so big that they are forced by nature to forge matter in their hearts. All the heavy atoms the Earth is made of, all the atoms that are necessary for life, atoms that your body contains, were once forged in the heart of a star. When you breathe, you inhale them. When you touch your skin, or someone else's, you touch stardust. You wondered earlier why stars like the Sun had to die and explode at the end of their lives, and here is our answer: without such endings, there would only be hydrogen and helium around. The matter we are made of would be for ever locked within eternal stars. The Earth would not have been born. Life as we know it would simply not exist.

To look at this another way, since we are not made out of hydrogen and helium alone, since our bodies and the Earth and everything that surrounds us also contain carbon and oxygen and other atoms, we know that our Sun is a second- or maybe even third-generation star. One or two generations of stars had to explode before their dust became the Sun, and the Earth, and us. So, what is it that triggered their death? Why are stars doomed to end their shining lives in a spectacular explosion?

One of the amazing properties of a nuclear fusion reaction is that however huge the amount of energy needed to start it in the first place – the weight of a whole star! – it then releases even *more* energy.

The reason for this may seem surprising, but since you see it happening right in front of your eyes, you have no choice but to accept it: when two atomic cores fuse into a larger one, some of their mass disappears. The fused core has less mass than the two that created it. It is as if mixing one pound of vanilla ice-cream with another pound of the same ice-cream did not give two pounds of ice-cream, but less.

In everyday life, that wouldn't happen. In the nuclear world, however, it happens all the time. Rather fortunately for us, the mass is not lost, though. It is turned into energy, and Einstein's famous $E = mc^2$ gives the exchange rate.[*]

In our daily lives, we are more used to exchange rates relating one currency to another rather than mass to energy. So, to see if $E = mc^2$ is a good deal for nature, imagine the same exchange rate is being offered at JFK airport to change pounds sterling (that's the initial mass) into US dollars (the energy one gets for it). The exchange rate is then c^2, where "c" stands for the speed of light, and "c^2" is the speed of light multiplied by itself. For one pound, you'd get 90 million billion dollars. A pretty good deal, I'd say. In fact, it is the best exchange rate in nature.

Obviously, the missing mass in each single nuclear-fusion reaction is rather small. But there are so many atoms fused every second within the Sun's heart that the energy released is enormous, and it has to go *somewhere*. So it pushes out into space, away from the star's core, in as many ways as is possible. In the end, the energy from this nuclear fusion balances the gravity that presses everything down into the core, making our star's size stable. Without it, were gravity the only player, the Sun would shrink.

Nuclear fusion emits a tremendous amount of light and particles, which happen to turn everything nearby into a shining soup of nuclei and electrons that is called *plasma*.

---

[*] I know you probably know, but let me say it anyway, just to make sure: in $E = mc^2$, "E" stands for energy and "m" for mass; "c," on the other hand, is the speed of light. So this equation, the only one you'll see in this book, means that one can literally turn mass into energy, and energy into mass.

This outburst of light and heat and energy is what makes stars shine.

The Sun, being a star, is not a big ball of fire – fire needs oxygen, and although the Sun creates bits of it along with other heavy elements, there's not enough free oxygen in outer space to sustain any fire whatsoever. A struck match would never catch fire in space. The Sun, like all the stars in the sky, is simply a bright ball of shining plasma, a hot mixture of electrons, of atoms stripped of *some* of their electrons (they are called *ions*), and of atoms stripped naked of all their electrons – the bare atomic nuclei.

As long as there are enough tiny nuclei to press together within its core, the Sun's gravity and fusion energy will remain in equilibrium, and we are lucky enough to be living near to a star that is in such a state.

Actually, it has nothing to do with luck.

Were our Sun *not* in that state, we would not be here.

And as you now know, the Sun won't be in that state of equilibrium for ever: our star's core will some day run dry of its atomic fuel. That day, there won't be any more outward push radiating from the core to compete with gravity. Gravity will then take over and trigger the final sequence of our star's life: the Sun will shrink and get denser, until a nuclear-fusion reaction is triggered again, but away from the core, closer to the surface. This reborn fusion reaction will not balance gravity, but overpower it, and the surface of the Sun will be pushed away, making our star grow. You saw it happening during your trip into the future. A final burst of energy will then herald the death you've already witnessed, spreading into space all the atoms the Sun has forged throughout its life while creating some more – the heaviest ones of all, such as gold. Eventually, these atoms will mingle with the remains of other nearby dying stars to form huge clouds of stardust that will, perhaps, seed other worlds in the far future.

It is by estimating the amount of hydrogen left in our star's core that scientists can guess when this explosion will happen, and the result says that the Sun will blow up in about 5 billion years from now, on a Thursday, give or take three days.

4 | *Our Cosmic Family*

What you've discovered about the Sun so far makes you more familiar with it than any human who lived before the middle of the twentieth century. All the light that showers your body day after day comes from atoms being forged in the heart of our star, from parts of their mass being transformed into energy. The Earth, however, is not the only celestial object that profits from the Sun's energy.

In the blink of an eye, your mind is back on the Sun's bubbling hot surface and you look around, like a hawk. Eight bright dots are moving against a seemingly fixed background of distant stars. These dots are planets, matter-filled spheres too small to ever dream of one day becoming a star. Four of them, the four closest to the sun, look like tiny rocky worlds. The furthest four are mostly made out of gas. They are still tiny with respect to the Sun, but they are giants compared to the Earth, the biggest of the four small rocky worlds. Apart from the Earth, however – and even though they were all born out of the same cloud of dust from long-dead stars – none of these worlds and none of their hundreds of moons are a potential shelter for humanity's future. They all are bound by the Sun's gravity, and they will all be gone with our star's ultimate boom. Shelter, if any is to be found, must lie further still.

With a feeling of urgency, your mind shoots as far away as is possible, to have a look at what lies beyond the Sun's sphere of influence. Along the way, you will pay a visit to our planet's distant cousins, the giants of our cosmic family.

You are about three times further away from the Sun than the Earth is. Mercury, Venus, Earth and Mars, the four small rocky worlds closest to the sun, are already behind you. From there, our star is a shining dot half the size of a penny held at arm's length. A typical July midday in the United Kingdom, the hottest day of the year, say, would feel colder than the coldest winter in Antarctica, were the Earth to be here.*

Sunlight gets scarcer and scarcer as you move away from our star.

You shoot past some rocks, leftovers from the early days of our planet's formation. They are mostly potato-shaped asteroids that, together, form what astronomers have come to call the *asteroid belt*, a huge ring of rocks that encircles the Sun, separating the four small terrestrial planets from a world of giants. The rocks themselves are pretty scattered and, as you fly through the belt, you realize there's hardly any chance of your hitting one of them. Many human-made satellites have flown through unhindered.

Leaving the belt behind, you now fly past Jupiter, Saturn, Uranus and Neptune, the gas giants, all enormous planets with relatively tiny rocky cores deeply hidden beneath huge, tumultuous atmospheres. All these planets seem to be blessed with a magnificent ring system, although Saturn's by far surpasses, in size and beauty, all the others combined.

You fly by them all and watch them with the respect such gigantic worlds deserve, even though they are not suited for life.

Beyond Neptune, the furthest planet orbiting the Sun, you may have expected to see nothing, but that is not the case at all. Another belt lies there, made of all sorts and sizes of dirty snowballs, again likely by-products of our solar system's birth, when its current members aggregated from the leftover dust of long-gone exploded stars. This belt is called the *Kuiper belt*. The Sun looks like a pinhead

---

* In 2013, one of NASA's weather satellites registered a temperature of −138.5°F in Antarctica: the coldest temperature ever recorded on Earth. Out where you are now, in space, it would be colder.

from out there, just another star. Hardly any warmth seems to reach these distant parts, but there is some action.

Every now and then, due to collisions or other perturbations, one or more of these dirty snowballs is expelled from its quiet and distant orbit around the Sun. Pushed towards our star, it slowly reaches warmer climes and begins to melt as it speeds against the solar radiation, leaving long tails of small icy rocks shining in the dark; it becomes one of those celestial wonders we call *comets*. The European Space Agency's sturdy Philae probe landed on one in November 2014, to study its surface. The Rosetta spacecraft that took it there orbited it as it approached and moved away from the Sun to watch its outermost layers turn into gas . . .

Poor Pluto – which recently got stripped of its planet title to be reclassified as a dwarf planet – is part of that icy belt too, together with (at least) two other dwarves, called Haumea and Makemake. It is funny to think that Pluto, with its moon Charon, is so far away from the Sun, and has so much space to travel to complete a single revolution of it, that less than one of its own years has passed between the moment it was discovered and called a planet, and the moment it got stripped of that title, seventy-six Earth years later. It indeed took astronomers decades to see that it was actually just a quarter of the size of our own Moon. The dirty-brown Pluto that you now fly by of course hasn't been affected in the slightest by its renaming, and you soon leave it behind too, heading further away still from the safe protection of our shining star.* Yet more dwarves cross your path, and more comets, and you even see frozen worlds that no living person has yet discovered, but your attention quickly turns entirely to a gigantic sphere that encompasses everything you've seen so far.

---

* NASA's New Horizon spacecraft reached Pluto in July 2015, to study it at close proximity, a historical first, which revealed extraordinary features no one had expected, including puzzling signs of rather recent surface activity.

All the planets, dwarf planets, asteroids and comets you've seen lie more or less on a flattened disc at whose centre shines the Sun. But what you are seeing now does not. A reservoir of billions of billions of billions of potential comets forms a huge spherical cloud that seems to occupy all the space there is between the Sun and the realm of other stars. This reservoir is called the *Oort cloud*.

Its size is astounding.

It marks the boundary of our star's realm, which contains all the members of our cosmic family, a family called the *solar system*.

Beyond, you enter uncharted territories and head for what you reckon is the star closest to ours. It was discovered in 1915. A century ago. Just when our universe began to be understood. Its name is *Proxima Centauri*.

5 | *Beyond the Sun*

Your body is still on a beach somewhere on our planet, but your mind is now as far from the Earth as the furthest manmade object has ever been.* As you crossed the edge of the Oort cloud, you exited the solar system and entered the realm of another star. As you crossed that fuzzy line – as if for you to truly understand what the boundary meant – you saw some of the solar system's outermost comets switch orbit: from a faraway curve centred on the Sun, their trajectory became a faraway curve centred on another star, the star you are now heading for, Proxima Centauri.

Proxima Centauri belongs to a family of stars called red dwarves. It is much smaller than the Sun (about one seventh its size and mass) and has a rather red hue. Hence the name. Red dwarves are very common, indeed scientists believe they account for most of the stars in the sky, even though they are too faint for our eyes to see.

As you get closer and closer to it, you continuously see it undergo violent changes in its brightness and expel huge amounts of burning hot matter in a rather erratic way.

Now, are there any planets around that angry red dwarf? As incredible as it may sound, you do see one. A planet orbits Proxima. Its discovery was made public on August 24, 2016, by a research team using the European Southern Observatory based in La Silla, in Chile.

---

* The furthest-travelled human-made object is NASA's space probe Voyager 1. Launched in 1977, it reached the outer boundary of the solar system in 2013. It is still transmitting data to Earth and able to respond to new commands. Its batteries are supposed to last until 2025. As of January 1, 2017, a signal sent from Voyager 1 takes about nineteen hours and seven minutes to reach the Earth at the speed of light. It will take longer in the future, as the probe shoots further still. For a live update on its position, you can check www.voyager.jpl.nasa.gov.

Any life form able to live there could plan for a very, very long future. When our star, the Sun, blows up, Proxima won't have changed a bit. As far as we know, it will still be shining the way it shines now for about 300 times the present age of the universe. A long time by any standards.

Being smaller than the Sun, the tiny atomic cores that make up Proxima are fused into bigger cores at a much, much slower pace. Size, starwise, does matter: the bigger the star, the shorter its lifespan . . . And for the planets that orbit them, distance is the key. To have liquid water on its surface (and be able to sustain life as we know it), a planet needs to be not too cold and not too hot. For that, it has to be neither too close nor too far from the star it orbits. The zone around a star that allows for liquid water to remain on the surface of a planet there is called the *Goldilocks zone*. Amazingly, the planet you are contemplating now is within Proxima's Goldilocks zone. *Proxima b*, as this planet is called, certainly wouldn't be as hospitable as the Earth, but it could last forever. And it may one day be colonized, to replace our gentle world . . .

Feeling somewhat guilty for having had such a thought, you turn around to look at your home solar system, at your home world, expecting the Sun to outshine all the other bright dots in the sky, but that is not the case at all, and the sheer size of cosmic distances suddenly hits you.

Were you not pure mind but instead a real space traveller, how long, you wonder, would it take to send a signal home from here?

Were you to be equipped with an interstellar mobile phone, you could have tried to call some friends of yours at each of your stops, to share your discoveries with them. Cellular phones transform your voice into a signal that travels at the speed of light, making communication on Earth seem instantaneous. In outer space, however, distances are usually too large, and nothing seems instantaneous any more. From the Moon, light takes about one second to reach the Earth. And another to return. Had you asked a friend on Earth if he could see you when you were up there, with binoculars, his answer would have reached you two seconds later.

From the Sun, it would have been worse. Light takes about eight minutes and twenty seconds to travel the distance between the Earth and the Sun. Conversations start to become tricky, since one must wait more than sixteen minutes between a question and an answer. But the Sun is still only next door in cosmic terms. A call dialled now, from where you are, near Proxima Centauri, would send a signal that would make a phone ring on Earth in about four years and two months. Any reply to a query of yours would take no less than eight years and four months to reach you.

You have still only got to the star second closest to the Earth after the Sun, but it feels a very long way from home, so you look for something to help locate yourself, so as not to get lost.

Remembering the beautiful Milky Way you saw from your tropical-island beach, you look around to see where its cloudy white patch of light now lies. To your great surprise, you immediately see that it does not appear as a thick straight line any more but like a tilted ring, with some parts brighter than others, and you somewhere inside it. You realize that if it looked like a streak from the Earth it was because the Earth itself, under your feet, was hiding most of it.

Without a second thought, having found no planet around Proxima Centauri, you head straight for the brightest part of the Milky Way.

You do not know it yet, but you are now travelling towards the centre of a gathering of about 300 billion stars. A gathering that has been called a *galaxy*.

When you think about it, there has to be something peculiar at the centre of a gathering 300 billion stars strong. Take the Earth. Its centre is the densest, hottest, harshest place there is (within the Earth). Take the solar system. At its centre lies the Sun, the densest, hottest, harshest place there is (within the solar system). This may not prove anything, but it hints that there is probably something big happening at the centre of a galaxy too. Something really big.

As fast as thought, you fly by tens of millions of stars. Some are much bigger than the Sun, doomed to live even shorter lives, and others are tiny, ready to shine their light for time beyond imagination. You also fly through stellar nurseries, dust clouds made out of the remains of hundreds of exploded stars, and stellar graveyards waiting to merge and become stellar nurseries. And now there you are. Near the galactic centre, whatever that means, and you suddenly stop.

Right in front of you, there is a ring again. A spinning, colourful ring made of scattered matter. Looking more closely, you see it is made of gas and billions of rocks and comets all moving around a thick, doughnut-shaped source of bright, energetic light.

What is happening here? What are these rocks and icy fragments that are moving around? You look around a little further, and what you see seems not to be possible . . . It is not only lost boulders that orbit that ring: there are stars too. Whole stars. Not planets. Stars themselves. And they are moving fast.

Until 2015, one of them was actually the fastest known object in the universe. It is called *S2*, or *Source 2*. From the Earth, scientists have seen it complete an orbit around the doughnut in about fifteen and a half years. Given the distances involved, that means it moves at an astonishing 11 million miles per hour. But how can that be?

What beast has enough gravitational power to keep such a lightning-fast object close by? Is it even *possible* to generate such a force?

Imagine a marble, and a salad bowl.

If you spin the marble too slowly against the wall of the salad bowl, it will immediately fall down to the bottom. If you spin it too fast, it will spiral up and fly out to break something in your kitchen. But if you spin it right, it will move around for a moment, in a circle, some distance between the bottom and the top, without escaping the bowl, without falling, until friction turns too much of its velocity into heat and slows it down.

Now, imagine that the marble is this super-fast S2 star and that there is an invisible bowl keeping it in orbit around whatever lies within the bright doughnut. In space, there is no friction, so there is no reason for the star to lose any of its energy.* From the speed of S2, we can thus imagine the shape of the bowl and hence what mass lies at the bottom.

This rather straightforward calculation has been made many times by scientists, and it always gives an incredible answer: to create a gravitational field with the right strength for S2 not to be hurled away into outer space, the mass of more than 4 million Suns is needed. That would be a huge star indeed.

But we have a problem: there is no visible star within the orbit of S2. You can look for one as much as you like, you won't find one.

From Earth, to see what this 4-million-solar-mass object that keeps S2 from flying away is, scientists have built telescopes able to detect specific light that our eyes cannot see, namely either ultraviolet light or, for a more impressive vista, the second-most energetic of all the lights we know, X-rays. Using such a telescope, they still

---

* For fellow scientists who are reading this, at this early point in the book I am neglecting the gravitational waves.

cannot see an object, but they do see energetic bursts of light originating within the ring, from some tiny location there. What keeps
S2 from shooting away not only isn't a star, it is also far from being
as big as it should be. So much so, in fact, that scientists have only
one explanation for what must be hiding there: a black hole. A
supermassive one.

Scientists have called it *Sagittarius A** (pronounced "A-star"), but
they can't really study it clearly from the Earth, its surroundings
being hidden by all the stars and dust and gas that lie between its
location and our planet.*

You, however, are right next to it, and if you wonder what triggers these bursts of energetic light that telescopes on Earth detect,
you are about to find out.

Understandably, though, being right next to an invisible monster, you don't feel very secure. Who knows what a black hole could
do? Could your mind get swallowed, possibly never to be reunited
with your body? Could it get stuck in there and doomed to wander
away from everything you know? Or could there be a hidden passage, a door leading to another universe, to another reality, as you
may sometimes have heard people say?

Uncertain what to do, you stare at the billions of tiny particles of
dust and other small rocks that make up the bright ring.

Less than a minute later, a huge potato-shaped asteroid flies past
you at 600,000 miles per hour. You watch it carefully. As it speeds
through the ring, you see it melt into tiny specks of molten matter,
burnt by the friction caused by the ring's dust. Just as a small rock
entering the Earth's atmosphere might become a shooting star and
burn up entirely without making it to our planet's surface, the

---

*   For history lovers, Sagittarius A* was actually first detected using a radio wave
telescope, in February 1974, by American astronomers Bruce Balick and Robert
Brown.

asteroid disappears long before it could reach whatever lies inside the doughnut.

As you turn around again to look for more action, it is not just a big piece of rock that you now see coming your way. It is a star. A big, shining, furious star. Like S2. But bigger still. Will it burn too? Will it get through? You see it dip towards its fate and fly through the doughnut at an angle. It is now within the ring and out of sight, but it promptly reappears, after completing half an orbit, in a strangely distorted way, as if a mirage caused by some weird force caused it to change shape. It continues to fly down. Tremendous stresses seem to be acting on it. Planet-sized chunks of the star are snatched off its surface. You try to stay calm, praying that there is nothing to fear, but you can't help it, and your thoughts suddenly feel weary, and heavy, preparing for a disaster of mind-bending proportions . . .

Until now, you were ethereal, oblivious to whatever forces were ruling the universe, but that is not the case any more. Loaded with heavy thoughts, you become subjected to gravity and you are in the presence of its master. Against your will, you are dragged inwards, you are being sucked in, as if sliding on an invisible yet slippery slope. You cross the ring of heated matter and get close to the in-falling star, now ripped apart, as it bursts into a blazing wing of white-hot plasma that spirals downwards, taking you in, towards the still invisible black hole.

Needless to say, your fears are all justified. Hundreds of billions of billions of tonnes of plasma are plunging in with you. Your heart beats like mad as you spiral down faster and faster and faster until . . . until a tremendous whirlwind force ejects you out. What remains of the star is transformed into extraordinarily powerful jets made of what seems to be matter turned into pure energy. Confused, you wonder if you've just slid into a parallel world inside a black hole, but you soon realize that you didn't, that you are

moving away from the monster, that you've been ejected, or
rejected, by the master of mass. The Milky Way's giant ring is now
apparent again, far away.

Like that marble made to spin too fast against the wall of a salad
bowl, you and the dust from the disintegrated star have been
expelled before having reached whatever the black hole is made
of . . . You fell in too fast and got slingshot away before reaching
the invisible monster, and so did the star, whose matter turned into
two jets of the most energetic types of light known to mankind:
X-rays and gamma rays. One is shooting up and the other down,
like two beacons aiming their lights not just to the gaping space
between the stars of the Milky Way, but further still, towards
greater voids.

The jets' speed is astounding, and so is yours. You are being carried away and fly by millions of stars, as if a giant finger wearing
the Milky Way as a ring was pointing to your destination.

Maybe it wasn't yet time for you to take the plunge into a black
hole. Maybe nature wanted you to see more of our universe's beauties before allowing you to travel within a black hole's deadly
grip . . .

Whatever the reason, your heart recovers and your thoughts
become light again, ridding your mind of gravity's grasp. You are
far away and have recovered the freedom to move as you please.
Still, you follow the jet for a moment, to see where it leads. And it
doesn't take you long to see that something strange is happening:
the surrounding stars seem to be less and less numerous. So much
so that soon there aren't any ahead of you any more. Some sources
of light are still shining far away in the distance, but they really are
much further than anything you've seen so far. Strangely, also, the
Milky Way ring is gone. Wondering where it went, you look down
and gasp at the most extraordinary vista you've ever seen. No
human being or man-made object has ever been blessed with such

a sight. Earth-based observations have sneaked a few images of the surrounding of the black hole you've just escaped, but not of this. Were you to call Earth from your present position, a reply – if any – would take more than 90,000 years to reach you.

You are above the Milky Way. Your galaxy.

If you thought, looking up at the night sky from your sandy beach, that it must spread all the way to the end of the universe, you now see that it doesn't. Far from being Everything, the Milky Way is but an island of stars lost in a dark immensity of a far larger scale.

7 | *The Milky Way*

The first men to have been to space all came back humbled by the beauty of our planet, and by its tiny size in an ocean of blackness. But that was just the beginning. What you are staring at now is more humbling still.

You knew the Milky Way was a galaxy, but only now do you see what that really means. From above (or below, it makes no difference), the whitish cloud in the Earth's night sky doesn't look like a cloud at all, but like a thick disc made of gas and dust and stars. Right underneath you, spreading over distances so large that light would take tens of thousands of years to cross, 300 billion stars, bound together by gravity, are spinning around a bright centre.

If you think of the solar system, with its planets and asteroids and comets, as our cosmic family, if you think of Proxima Centauri as our neighbouring star, then the Milky Way can be thought of as our cosmic megalopolis, a thriving city 300 billion stars strong, the Sun being just one of them.

Intertwined within a whirling dance, surrounded by emptiness, it is such gatherings of stars and dust and gas that scientists call *galaxies*. And just as our star has been named the Sun, the Milky Way is the name we have given this particular galaxy, our galaxy.

Four huge bright spiral arms separated by dark patches swirl around its centre, where they meet up in an even brighter bulge of gas and dust and stars that hides everything down to the black hole you've just escaped. Only the jet of energetic matter being expelled from it, the jet you've travelled with, is visible from where you are.

If you have a hard time grasping what 300 billion stars floating on their own actually means, do not worry too much about it: nobody really can. But if you were to try to explain what you are seeing from up there to your friends once back on your tropical island, numbers won't help. Instead, tell them to pick a cubic cardboard box one metre high and fill it to the top with coarse sand from your beach. Now ask them to fill 300 such boxes with the same sand. There are as many stars in our galaxy as there are grains of sand in all these boxes taken together. Kindly ask your friends to fly back to New York City and pour the contents of these 300 boxes in a disc shape covering Liberty Island, around the Statue of Liberty, and to draw four spiral arms on it. Then tell them to climb to the top of the Statue of Liberty and look down. That's what the 300 billion stars of the Milky Way look like to you now. And if you tell your friends that you marked one of the grains of sand with a yellow dot before they winched themselves up, ask them to figure out which one it is. They'll realize how hard a time your mind is having up there, above the real Milky Way, figuring out where the Sun is. Not to mention the Earth, which is a hundred times smaller. Finding a star is hard, but planet hunters have a harder job still.

From above the Milky Way, to find the Sun, your mind does however have an advantage over your friends: you can imagine all the night-sky pictures that have ever been taken by humans, from Earth and from space alike, to compare them with what you now see. Over the years, scientists have created a map of our galaxy's stars, and without ever leaving the Milky Way, they have a fairly precise idea of where the Sun (and the Earth) lie within it.

To match the pictures of the night sky, you first concentrate your efforts near the galactic centre, near the bulge and the black hole, where all is bright and beautiful and powerful. Wouldn't it be natural for a species as important as ours to have blossomed in, or quite near, that very special position? Wouldn't it be logical, given our importance, and so very right, for the Sun and the Earth to be part of that galactic magnificence?

Well, they are not. The solar system lies about two thirds of the way between the central black hole and the outskirts of our galaxy, somewhere on one of the four bright arms. Not a special place whatsoever.* And to further rub salt in the wound, as you shall now witness, however huge it may be compared to us, even our galaxy itself is pretty insignificant on the cosmic scale.

Turning around to face whatever is left out there to see beyond the Milky Way, you catch sight of some shining blobs that seem to light up the far distant universe. You wonder: are they loose stars, these blobs? They seem a bit too smeared out . . . And far away . . . Are they . . . Could they be galaxies too? Can you see them from the Earth with the naked eye?

The answer to the latter question is no.†

On Earth, every time you've raised your eyes towards the night sky, all the stars you have ever glimpsed a twinkle of belonged (and still belong) to the Milky Way galaxy, to the spiral disc you've just seen. All of them. Even those stars that seem rather far away from the whitish band that streaks the night skies. The Milky Way is not an infinite sphere but a finite disc, and the Earth happens not to be at its centre, but rather closer to its edge. Different directions in the sky hence look very differently filled with stars, just as the night sky is different seen from different places on Earth: each place faces a different part of the Milky Way.

It so happens that the Earth's axis is tilted in such a way that the southern hemisphere always faces the galactic centre while the northern hemisphere always looks away from it, where there are far fewer stars. Accordingly, nights in the north are rather dull compared to southern ones.

From your tropical-island beach, what you called the Milky Way was just a slice of your galaxy, a band containing hundreds of millions of stars too far away to be seen individually, but whose lights,

---

*   Our existence may make it so, though.

†   Unless you have very, very good eyes and know where to look.

together, formed the cloudy band. As you now peer into the far-away unknown, ready to make your mind bounce towards whichever place you find most mysterious, you suddenly realize that all these blobs of light look as fuzzy as the Milky Way.

They must be galaxies, too. With their supermassive black holes.

And as you think so, right there, at an angle, another galaxy suddenly rises. The sight is astounding. Its edge appears from underneath the Milky Way and it is now growing fast. It is *Andromeda*, our local galactic big brother. It is so big that it is hardly believable that it took so long for mankind to discover what it was.

From Earth, Andromeda spans a part of the night sky that is about six times as large as a full Moon, but it is so far away that despite its 1,000 billion stars, only its central bulge can be spotted with the naked eye. And that bulge is tiny. The first human to notice it (and whose written records reached us) was the extraordinary Persian astronomer Abd Al-Rahman Al-Sufi. Around the end of the first millennium, more than a thousand years ago, when many, throughout the world, spent their short lives fighting each other, inventing cunning torture devices and fearing the end of the world, he watched the stars. Al-Sufi was one of the greatest astronomers of the golden age of Baghdad but, describing Andromeda's central bulge as a faint cloud of light, he had no way of knowing that it was another galaxy. He did not even know what a galaxy was. Actually, that knowledge came about a thousand years later. No one knew about galaxies as separate gatherings of stars until the 1920s and the observational works of Estonian astronomer Ernst Öpik and US astronomer Edwin Hubble. They were the first to notice that there were great expanses separating these other groups of stars from the Milky Way, making them separate entities in their own right.*

---

* Some had thought about that possibility before them though, the first of whom seems to be the eighteenth-century English astronomer and mathematician Thomas

Andromeda is the closest cosmic proof that the Milky Way is not the entire universe.

As you look at it, as you realize that the Milky Way and this majestic spiral of 1,000 billion stars whirl around each other, you also become aware that all the galaxies in the universe are engaged in a cosmic ballet, a ballet where the dancers are lone shining islands, gatherings of billions of stars moving within the dark emptiness of space.

Spanning the cosmic horizon, an extraordinarily powerful feeling crosses your mind as it starts to encompass the Milky Way, and Andromeda, and other galaxies both near and far.

In a pure, blessed moment, you suddenly see everything. Tens, hundreds, thousands, millions, hundreds of millions of galaxies. Everywhere, forming groups of various sizes. Forming weird, filament-like structures that crisscross the whole visible universe.

Who would have thought?

A few minutes ago – or is it hours? – you were lying on a beach, on holiday, and now the entire visible universe is contained within your mind. You've reached such a viewpoint that the dots studding the universe are not lone stars any more, but groups of galaxies, each group containing thousands of galaxies, themselves made out of hundreds or thousands of millions of stars, the Milky Way being but one of them.

As you embrace this amazing picture, as you look at all these places, you can't help thinking that you'd have the same trouble finding your home galaxy out of all the others as you'd have finding the Sun within the Milky Way, or a grain of sand on Liberty Island. Still, you set your mind free and shoot at the speed of

---

Wright. A few years later, the German philosopher Immanuel Kant elaborated on his idea.

thought and see galaxies spin and dance and whirl, and get torn apart, and smash against one another, and you witness tiny ones disappear as they purely and simply are swallowed by some giant neighbour.

Now wait.

Should you worry about that?

In a blink, you are back near the Milky Way. Andromeda is overhead. It is huge. Could it be that it, too, will one day merge with the Milky Way? The two galaxies are certainly moving around each other but something else is going on . . . Sharpening your gaze, you suddenly jump as you realize that Andromeda and the Milky Way really are falling upon each other, at the astonishing rate of 62 miles per second, leaving just 4 billion years before they collide.

They will start to merge 1 billion years before the Sun explodes.

Gulping painfully, wondering how humanity could ever be saved from *that*, a note of relief crosses your mind: galaxies are so big, and there is so much space in between any stars within, that galactic collisions hardly ever make stars smash into one another . . . There is a risk, sure, but you'll have to live with it for now.

It is absolutely normal, at this stage, to go through a Copernican Philosophical Depression. You may even begin to wish you were still living a few millennia ago, when the Earth was flat and, for the obvious reason that we humans like to consider ourselves special, at the centre of the universe. How reassuring it must have been to believe that everything revolved around us, that angels were spinning some sacred wheels attached to a cosmic clockwork mechanism, making the stars and the Sun move! Why on Earth did the fifteenth-century Polish mathematician and astronomer Copernicus have to ruin all this and proclaim that the Sun was *not* orbiting the Earth? Why did the seventeenth-century mathematician and astronomer Galileo see that Jupiter had moons that were *not* orbiting the Earth (or the Sun, for that matter, since they are orbiting Jupiter itself)? Why did Öpik and Hubble see that there were other galaxies out there? Why? They started all this!

Well, besides the fact that they were right, without the likes of Copernicus and Galileo and so many others, humanity would be doomed and – which is arguably worse – I would never have written this book. You would never have travelled by thought alone through our cosmic vicinity, not to mention beyond (as you are about to do). And between you and me, wouldn't it be a shame if all the beauty that is hidden out there were left unseen, or unexplored or – worse still – left only for other intelligent species to see from their own faraway cosmic perspective?*

And again, while we're on the subject, and as the sheer size of the visible universe starts to sink into your mind, do other species actually exist? Within those billions upon billions of star groups that stud an otherwise dark universe, are there red dwarves like Proxima Centauri circled by planets of their own? Are there double-sun systems shining upon inhabited worlds? Are there other Earths?

It may seem scarcely possible to believe we're alone in this gigantic universe: "If it's just us, seems like an awful waste of space," wrote US astronomer and cosmologist Carl Sagan in 1985 and yet, thirty years later, no one from Earth knows. The existence of alien life is still an exciting possibility (and a scary one too, granted), but for now it's just that: a possibility. This may change very soon, however, as our telescopes start discovering more and more worlds out there. I, for one, very much hope so anyway.

Even during some of the darkest years of humanity's chaotic past, some people heroically defied the religious authorities by asserting

---

* I have written *"other* intelligent species" here, but as English theoretical physicist and cosmologist Stephen Hawking often jokes (or does he?), we've still to find proof of intelligence here on Earth.

that, indeed, other worlds probably existed. The Italian Catholic monk Giordano Bruno, for one, got burnt alive in Rome in 1600, for having dared to speak such a heretical thought out loud: he had claimed that there were "countless Suns and countless Earths all orbiting around their Suns." He died in agony for believing so.

Today, even if, in my opinion, far too many people (even in the most developed countries) would rather play deaf and blind than face some facts unravelled by science, we know better than any Inquisition. Potential Earthlike planets have been found, and the likes of Giordano Bruno have been vindicated many times over, albeit rather recently.

Humanity has known about the existence of planets like Jupiter or Venus for ages, that is true. But the first time in history someone actually saw a planet orbiting a star that is *not* the Sun was only about twenty years ago, when, in 1995, two Swiss astronomers, Michel Mayor and Didier Queloz, spotted a giant world, which they called *51 Pegasi b*, orbiting a star located about 60 light years away from us.

The planet Mayor and Queloz found is not habitable, though, if only because it is far too close to its star. Still, it is a planet. Following their discovery, a few other such worlds were found each month, until specially devised satellites were sent to find more. NASA's Kepler telescope, launched in 2009, is one of them. Today, more than 6,000 candidate worlds have been detected. Out of these, about 2,000 have been confirmed as being planets orbiting faraway stars. Some are even double-star systems (planets orbiting two suns), and many other surprises are sure to break future news. To differentiate them from Venus or Jupiter and the other planets that are part of our sun's family, these faraway worlds are all called *exoplanets*. And by the way, out of the 2,000 confirmed exoplanets, about a dozen are potentially Earthlike and at least three of them, including one whose existence was confirmed in 2015, does bear stunning similarities with our home (the 2015 one is called Kepler 442b) . . .

All of these other worlds may be barren, of course – but they may also harbour life. In fact, I am ready to bet that direct or indirect signs of extraterrestrial life will be found within the next two decades or so. Maybe on one of these candidates, or on some yet-to-be-discovered ones. The technology is almost there for us to detect signs of biological activity within the atmosphere of such remote worlds. It would be great to live through such a discovery, wouldn't it?

Now, all the exoplanets that have been detected so far are within the Milky Way, our galaxy, and thus rather near to Earth. Planets that could exist in *other* galaxies are much too far away for any of our telescopes to spot, even though there may be hundreds of billions of them out there.

The Andromeda Galaxy, for one, may very well be teeming with life. It is the largest of all the galaxies that surround ours. And it is very close. On a galactic scale, that is. Not on a human one. A call made right now from Earth to some place around one of its 1,000 billion stars would take about 2.5 million years to reach its destination. If we were to make contact, we'd better find an intelligent question to ask. And an appropriate language too.

8 | The First Wall at the End of the Universe

Now, how big is the visible universe?

What would happen if you were to shoot straight out into what can be seen, for as long as is possible?

Is there a limit to it all?

Well, since someone is, sooner or later, bound to ask you that once you are reunited with your body, you'd better try to figure it out.

With confidence, you pick a random direction and head for it.

As you start moving away from your home galaxy, you immediately realize that the Milky Way is part of a little group made of fifty-four galaxies gravitationally bound to each other. Scientists have called that group the *Local Group*. It spans a sphere about 8.4 million light years wide. The Milky Way is its second-largest member, Andromeda being the king.

Beyond lie other groups of galaxies. Some of them are several hundred galaxies strong. These big gatherings, much larger than ours, are called *clusters* of galaxies. As you keep shooting ahead, you fly by giant clusters, *superclusters*, containing tens of thousands of shining spirals and oval discs made of countless stars, all bound to each other by gravity and stretched throughout space and time.

These superclusters form mind-bogglingly large structures.

As you shoot away from everything you know and look at the universe on a different scale, you realize that you will once again have to reconsider your relative size in the grand scheme of things. Your mind's eye wide open, you turn around, and you look everywhere, gathering as much light as is possible from all possible directions, searching for an end to all this. There is no notion of up or down, no difference between left and right. You are now more than 1,000 million light years away from the Earth and billions upon billions of shining galaxies are spread throughout an unbelievably large darkness. Around you, near and far, galaxies and groups of galaxies and clusters and superclusters are separated by distances larger still, larger even than all the distance you've travelled so far.

That the Milky Way is but one of all these dots is barely believable, and yet you know that what you see is no fantasy, but what is known to mankind.

Still, true facts or not, the idea of saving the Earth no longer seems to make any sense at all. Why bother? Why care? Understandably, leaving everything behind and letting yourself forever drift in this beyond-beautiful hugeness of a reality even becomes an appealing dream. Why not spend your life up there? Is that what scientists do, daydreaming in their laboratories?

As you contemplate the idea of never returning to your daily life, a strange feeling catches you and begins to inject a new energy back into your mind: somehow, all you are now seeing, all you are now moving through, is what mankind understands the universe to be. Somehow, you are travelling through the universe as it is pictured by human minds, so all this immensity has to be contained within the limits, if there are any, of a human brain. As amazing as it may sound, it is a reassuring idea and it brings you back to being a human being, a member of a species able to project its thoughts as far as the eye can see, and far, far beyond . . . Embracing the spacescape, you wonder: could the magnitude get still bigger than this? Could your mind embrace even more? Whatever the fate of the Earth, you decide that you'd rather know than not. Your virtual heart thumping with renewed curiosity, you desperately head forward and fly past thousands of millions more galaxies. As is invariably the case with humans, familiarity soon sets in, and even the universe's immensity ceases to shock you. What may have felt like desperation a second ago now seems to have turned into joy.

Here and there, you see galaxies colliding, you see stars burst into superstars, *supernovae*, outshining billions of their siblings for but the wink of an eye. Throughout the universe, everything moves around everything, you are being blessed by a show of amazing proportions and inhuman beauties.

Heading forward without looking back, you are now 10 billion light years away from Earth.

Your mind keeps flying ahead and away.

You are 11 billion light years from the Earth.

Twelve.

Thirteen billion light years away, and counting.

And you now feel elated, and you look for the end of our universe, and you see none, but your mind slows down slightly, for the galaxies around are getting scarcer. And the stars they are made of seem to be getting bigger. Enormously so, in fact. Some of the stars you now see are hundreds of times bigger than the average stars of today's Milky Way. You keep moving forward, albeit at a slower pace. The number of shining sources of light in front of you has now fallen drastically. And as you reach a distance of about 13,500 million light years away from Earth, pretty much all sources of light are gone.

You stop. Could you possibly have reached what you were looking for? Does the universe actually have an end?

You remember having raised the question a couple of times with your friends prior to your tropical-island trip together, but you had never given the thought a real meaning. And now you wonder, did you think you could eternally zoom out from Earth into the outer universe and continue to see galaxies?

Well, since you are travelling through the universe as it is seen from Earth, let me say this: our telescopes have shown us different. There is indeed a limit to what we can see, and will ever be able to see, using light. Your mind hasn't reached that limit yet, but it soon will. For now, it is travelling through a place so remote in space and time that the first stars weren't even born. For this reason, the place and epoch you are crossing has been dubbed the *cosmic Dark Ages*. Any light we see coming from there has travelled throughout the universe for 13.5 billion years to reach us. It is back then, in a stretch of time lasting about 800 million years, that the first stars began their job of transforming small hydrogen and helium atoms into the matter we, and other planets and stars, are made of. These were the first generation of stars, our Sun being a second- or third-generation one.

•

As you continue to fly forward, expecting the darkness to prevail for ever, you suddenly reach a place through which light cannot travel any more.

The surface of what seems to be a wall in space and time.

Beyond, the universe is not dark. It is opaque. You stop right in front of it and cast a virtual hand forward, to gently probe what lies beyond.

Goosebumps spread over your non-existing flesh as you touch what seems to be a tremendous amount of energy. An energy so dense that you suddenly understand why light cannot travel there: it would be rather like lighting a torch within a wall. Light exists beyond the surface you are facing, but it has no freedom to travel at all.

What you have reached is not a product of your imagination. It is the furthest place our telescopes can see. It is the place in space and time where and when our universe became transparent. No light from beyond, no light from before that time, will ever reach the Earth in a straight line. No light from before that place will ever be caught by any of our telescopes.

It took many decades for theoretical physicists to understand what this meant. In the end, as you shall see in the next chapter, they came up with a rather bright idea to make sense of it all, and this idea is called the *Big Bang theory*.

For now, however, you'll have to accept that you just reached the end of the visible universe. It is a surface that has been detected, and mapped, with our telescopes. The surface of a wall that no light can cross. It has been called the *surface of last scattering*.

But just as you start realizing how weird and unexpected all this sounds, everything disappears around you and you find yourself back on your tropical-island beach, looking up at the night sky. The stars are still there, and so are the trees and the sea. And so are your friends. They are looking at you in a rather peculiar way.

You sit up and tell them the extraordinary journey you've just been on. The Sun dying – we *need* to find a solution to that problem – the universe so big it's crazy . . . and the wall! Out there, the wall that marks the passage from opacity to the Dark Ages!

Your friends' peculiar looks turn into worried ones. As they help you up and walk you back to your villa, you can hear them wonder if perhaps the barbecued shrimps were not fresh, or the alcohol too strong.

To the east, a few hours later, some of the rising sun's rays begin to bounce off the dust contained in the Earth's atmosphere (especially the ones corresponding to the colour blue), diffusing all around, hiding space away from view. Lying down on your bed, surrounded by early morning birdsong, you open your eyes and see the silhouette of one of your friends standing right next to you. She apparently kept watch over you throughout the night. Did you dream it all? you again wonder. Did your mind really travel through the vastness of space?

As your friend asks you if you are feeling better, holding a glass of water out for you, a fresh morning breeze gently caresses your forehead, and you smile, thinking that either way it feels good to be back on Earth.

And then you smile even more, for deep inside yourself, you know that you experienced something very special and dreamt none of it, that it was all true, that you were blessed with *seeing* without having to study for years. For some reason unknown to you, you saw the universe as it is known today.

Relieved to see you smile, your friend stands up to fetch you some breakfast. As soon as she is gone, however, you immediately try to recall what you've been through, not to forget, with the feeling that this was only the beginning of a very strange adventure.

Sitting on your bed of woven palm leaves, watching the waves wash over the seashore, you remember the Earth as seen from space, a tiny dot of blue orbiting the Sun. You remember the other stars, billions of them, swirling around the central black hole that hides near the centre of the Milky Way, our galaxy. Then you remember Andromeda and the four dozen or so galaxies that make up the Local Group, and then you remember the other groups and galaxy clusters and superclusters that spread far, to infinity and beyond.

No.

Not to infinity.

To the Dark Ages and the wall. The surface of last scattering, beyond which no light can travel freely.

And you know that whatever direction your mind might have taken on its journey, you'd have ended up hitting that wall.

This pretty much sounds as if, on a scale much larger than

anyone could ever imagine, the Earth was in the middle of a sphere, a sphere whose boundary is made of that wall. What lies within that sphere may well be the entire visible universe that will ever be accessible to mankind.

As you let that thought sink in, you blankly stare right in front of you, at the horizon.

If the surface of last scattering surrounds the Earth, then the Earth *has* to be at the centre of a sphere bounded by that wall.

That sounds logical.

But then, that means the Earth *is* at the centre of its visible universe.

Shocked, hardly believing it yourself, you shake your head and mutter that none of this makes sense.

It makes no sense at all.

Still, you know what you saw, and you suddenly wish you could get back up there and have another look at it all.

Which you will, but from a different perspective, very soon.

To prepare yourself, let me just say that the surface you saw, the surface of last scattering, is not the end of the story. There are at least two other surfaces beyond, with walls behind them. The first one is called the Big Bang itself. The second one hides what *caused* the Big Bang.

Before you reach the end of this book, you'll travel all the way to that second wall, and beyond.

But first, you should take it easy.

You are on holiday, after all, and your friend is back with your breakfast.

While you eat, however, I'll help you bring some order to what you've been through.

Part Two

Making Sense of Outer Space

1 | *Law and Order*

Have you ever tried to jump off a cliff? Or through a window from the top floor of a skyscraper?

Probably not.

Why?

You'd be dead.

And so would I, if I were to try it, and so would anyone.

Now, why do we all know this?

The answer is as straightforward as it is mysterious and deep. In it lies the reason why the human race has already managed to conquer the Earth and a small portion of the sky. In it lies the reason why we managed to send you to look at the stars in the first part of this book. It has to do with nature and its laws.

However learned we are, whether we liked science at school or not, whether we are scientists or not, if we search deep within ourselves we all have the intuition that there are laws in nature, and that these laws cannot be broken. That anyone jumping from too far up is doomed to fall towards a splashy end is one of them.

Over the millennia that separate us from our hunter-gatherer ancestors, many men and women have continuously been searching for these laws. And they managed to find some, too. Today, the field that endeavours to continue this quest and further unravel the mysteries of nature is called *theoretical physics*, and it is its (always under construction) kingdom whose doors are about to open for you to travel through.

This kingdom was arguably built when English astronomer, physicist, mathematician and natural philosopher Isaac Newton created a new language – that of mathematical analysis – which allowed him to describe pretty much everything that is within the reach of human senses. Why, when a person steps from a cliff, he or she falls rather than walking on air, is given by a formula. As long as we know how the fall starts, Newton's formula tells us where and how fast the fall will end. The same formula also says that there is no difference, when falling from a cliff, between a human, a sponge or a piece of rock, as long as we forget about the drag of air-induced friction. It also says that the Moon completes an orbit around the Earth in a little under twenty-eight days, and that the Earth revolves around the Sun once a year. That particular formula is called *Newton's universal law of gravitation*. For this, Isaac Newton is still considered today as one of the greatest minds of all time.

There is no need to be a scientist to guess that finding such a law must have felt very good and Newton must have been very pleased with himself indeed. Strangely, however, instead of throwing parties every night to celebrate (as I would have), he preferred to make sure he was right. So he started to check if his gravitation formula really deserved to be called universal. Scale is of the essence here, because as you've already figured out in the first part of this book, compared to the universe, the Earth isn't much to shout about, to say the least. And what is true for a tiny speck of dust may not be true for a galaxy.

On Earth at Newton's time there was not a single experiment that could prove his formula wrong or even challenge it. An arrow, for instance, always landed where it should. And a mountain would have too, had someone thrown one.

Now, what about even bigger things? What about places where gravity's effects are more intense than those found on our planet? To figure that out, we have to look beyond the Earth. And since you've already been touring the nearby universe, you know that the most obvious and easiest place to start checking is also the brightest, the Sun.

2 | *A Troublesome Piece of Rock*

Our star's surface gravity – the way it pulls you down on the surface – is about twenty-eight times stronger than our planet's, but the Sun is not the most gravitationally powerful object you encountered while discovering outer space in the previous part of this book. Black holes, for one, are much more potent. Still, the Sun outclasses the Earth. And it is much easier to probe than black holes. So, does Newton's formula work around our star as well as it does around our planet? And how could we check?

As you've seen, there are eight planets in the solar system. From the furthest from the Sun to the closest, these include Neptune, Uranus, Saturn, Jupiter, Mars, Earth and Venus. Maybe we could have a closer look at how they shoot through space, and check whether or not the Sun pulls them towards itself the way Newton's law says it should. Thanks to many astronomers who put aside their family life to look at the stars throughout their nocturnal existence, humanity even had a precise description of some of these orbits by Newton's time.* And the answer is almost too good to be true: if one takes into account how the planets interact between themselves too, all the planets listed above† move precisely according to Newton's formula. What a relief . . . The formula really is universal after all. Newton's mother must have been very proud.

But wait, hold on. Those of you with sharp eyes will no doubt have noticed that one planet is missing in the list above. We've only named seven of the eight planets that belong to the solar system. We've forgotten one. The one closest to the Sun. The one that feels the Sun's gravitational pull more than all the others. Mercury.

---

* Uranus and Neptune were discovered later. Thanks to Newton's formula, actually.

† Including Uranus and Neptune.

And with Mercury, there is a tiny problem. A slight mismatch. Nothing big. Something so small that, surely, it can't matter much. But it does. Over the couple of centuries that followed Newton's work, this slight mismatch changed everything humanity knew about space and time.

Mercury isn't that impressive. Only slightly larger than our moon, it is the smallest planet of the solar system. It is rocky and its surface is battered with craters that are unlikely to disappear anytime soon. Mercury has no atmosphere, no weather to smooth out irregular shapes and scars. In short, Mercury is not the kind of planet one would choose as a holiday destination. To complete a full spin on her axis takes her fifty-nine Earth days, meaning that a night on Mercury lasts a month on Earth, and is followed by a similarly long day. Both day and night on Mercury are hellish. Daytime temperatures can reach 806°F only to drop at night to −292°F. Newton did not know these details, and he probably couldn't even guess how harsh a world Mercury is. Today, we know. And today, we also know that according to his formula, the trajectory of all the planets around the Sun should look like a slightly squashed circle. As I said above, for all the planets, Newton's calculation was (and still is) in perfect agreement with observations. Were they to leave a trail behind them, each of the planets would draw a squashed circle, an *ellipse*, in the sky, a path they would pretty much retrace year after year, just like Newton said they should. But not Mercury. Mercury's orbit happens to spin on itself, like an egg tumbling end over end, so that Mercury does not retrace the same path twice. This is *mostly* due to the other planets – they pull tiny Mercury towards themselves every time they get close – as Newton had already guessed. *Mostly*, though. Not entirely. The mismatch is tiny, but it is there. Visualize the space between two consecutive seconds on a watch (an old-fashioned one, with a big hand and a little hand) and divide that space by five hundred. A single one of those divisions is the angle by which Mercury's squashed circle drifts away from Newton's calculation over a century.

It may seem unbelievable that such a tiny drift could ever be identified without scientists having to wait some hundreds of thousands of years, but it was. Worse: we now know there is no way Newton's formula could predict it, let alone explain it, because that mismatch has to do with an aspect of gravity far beyond whatever Newton could have imagined.

Newton's equation quantifies how objects attract each other, but it says nothing about what gravity actually *is*. Poor Isaac (and many other scientists) actually spent a rather large amount of time trying to understand where gravity came from. Is it a property of matter itself that makes objects attract each other? Are all objects in the universe linked? And if so, by what? No visible or invisible elastic rope has ever been detected between our feet and the ground of our planet, or between the Earth and the Moon. And what about a magnetic link? Well, magnets do not stick to our feet when we try to hold them there, because our bodies are electrically neutral. Gravity can't be a magnetic force, then. So what is gravity? And why does stubborn Mercury, the tiniest of planets, beg to differ from the others?

Newton died in 1727. He had failed to find an explanation. One hundred and eighty-eight years passed before someone suddenly came up with a rather strange new idea.

The good thing about research in physics is that when observations fail to be consistent with theory, the first thing we say is that the observation must be wrong. Then we try and do the experiment again, and when experiments stubbornly and repeatedly give a wrong answer, we check if, by any chance, someone unknown had predicted such an outcome using an alternative theory. If the answer turns out to be "no," it is fair to assume we have no idea whatsoever why nature behaves this way. The safest option is then to try everything out. Obviously "everything" includes the craziest ideas, and that, I have to say, is a lot of fun. As we shall see later, ideas that are being probed today to figure out how our universe was born are worthy of the best science-fiction models (and as the Astronomer Royal, Sir Martin Rees, Baron Rees of Ludlow, once said, good science-fiction is better than bad science). In general, of course, most of these ideas are completely wrong. But never mind. What is important is to search and see what happens. So far, this approach has worked rather well.

Newton's formula had thus been used for nearly two centuries without any problems, and, to be fair, Mercury's case did not have much impact on most people's lives. But then, a scientist came up with a completely deranged idea about gravity.

Imagine the Sun, in space, with Mercury spinning around it, and forget about everything else. They are alone in the universe. A small rocky planet orbiting a huge shiny Sun. Emptiness all around.

Now get rid of Mercury. And get rid of the Sun too.

(Just to be clear: there should be nothing left.)

What if gravity had something to do with that "nothing" that is left, i.e. with the very fabric of the universe (whatever that may be)?

To figure out what could happen if this was the case, let's put the

Sun back and think. If we assume for a moment that the fabric of our universe can be moulded, one of the simplest interactions the Sun could have with it is to bend it. How might that happen? Well, try to picture a heavy ball lying on a stretched rubber sheet. The rubber will bend downwards around the ball. If you then cover the rubber sheet with soap, anything walking on it – an ant, say – that gets too close to the bent part will slip towards the ball, downwards. For the ant, that effect could be felt as gravity.

Obviously, if the stars and the planets were all lying on a soapy rubber cloth, I should hope we'd have noticed by now. So the fabric of the universe can't really be a flat, solid rubber sheet. However, it might be a 3D invisible one, or even a 4D one. And whatever that voluminous fabric were made of, why not imagine it bending around the matter it contains? Not just along a plane, of course, but in all directions, as if a ball immersed in the ocean were bending the water around it.

Taking this idea seriously for a moment, gravity would then merely be the result of that bending: whenever one falls, one falls not because of a force that pulls things down, but because one slips down an invisible slope in the fabric of the universe (until one hits a ground of some sort which prevents further falling).

A crazy idea, yes, but why not, after all, give it a try? How might things move in a universe like this?

For all the planets up to Mercury, geometrical calculations using this "bending" theory actually give the exact same results as Newton's. Which is as reassuring as it is exciting. So what about Mercury?

The man who came up with this deranged idea of a "bending" found that, in a universe like the one he described, the squashed circle of Mercury's orbit should spin around the sun in a way that does not agree with Newton's calculation. By how much? By an angle corresponding to about one 500th of a second on a watch.

Every century. Amazing. For more than fifteen decades after Newton's death, nobody was able to figure that out. But he had. He was
right. Gravity suddenly wasn't a mystery any more. Gravity was a
bending of the fabric of the universe caused by the objects it contains. Newton had not seen this. No one had seen this before, and
we are still trying to figure out all the consequences of this vision
today.

Stephen Hawking has often said: "I wouldn't compare the joy of
discovery to sex, but it lasts longer." A single look at the picture of
the man who solved Mercury's problem seems to confirm that
statement.

His name is Albert Einstein and the theory we've just introduced, the theory that links matter and the local geometry of the
universe into a theory of gravity, is called the *general theory of relativity.*

This theory was first published in 1915, a century ago, and scientists took some time to realize that Einstein had incidentally
revolutionized our vision of everything. Contrary to what everyone had believed before him, he had found that our universe not
only could have a shape, but that it was dynamic, i.e. able to change
with time. As stars and planets and everything there is move, the

bending they create in the fabric of our universe moves with them. And what is true locally around these objects may well also be true for the universe as a whole. In other words, even though he did not believe it himself, Einstein had found that our universe could change over time; that it could have a future. And if something has a future, it could also have a past, a history, and maybe even a beginning.

Before Einstein, it was understood that our universe had always just *been*. Now we know that it hasn't, at least not in the way we experience it. And we've now known this for a hundred years. So, as far as knowledge is concerned, the universe we live in, our universe, is a hundred years old.

Travelling through the known universe as you did in Part One is a bit like strolling around a forest on your tropical island and being amazed by the beauty of the trees. After such a walk, you could of course go back to your villa, have your friends over for a drink and tell everyone how lovely it is out there, how good it felt to breathe the fresh oceanic air. But then, your friends could rightfully wonder why the trees grow, why their leaves are green, how all these plants came to be the way they are . . .

If the universe is our forest, what is there to figure out about it? Instead of questioning the freshness of the shrimps you ate, what should those friends of yours have asked regarding the big picture? Beyond looking at it, is there anything to understand at all? And, seriously, is it actually possible to travel out there the way you did?

To that last question, the answer is easy: with your body alone or in a spaceship, no. As far as we know – for now – it is not possible to travel through space and time like this any other way than in your mind. Nothing that carries any kind of information whatsoever is able to travel at a speed greater than the speed of light. So what your mind did in Part One was to actually fly through a frozen 3D picture of the universe as it is known today, a reconstruction obtained by patching together all the pictures taken by all the telescopes that have ever been built on Earth. You may retort that you saw things move, that it wasn't a still picture . . . Fair enough. So let's say it was an "almost frozen" picture. Now, what can we make of it? Is there any law that governs the evolution of everything?

The morning after your mind trip, when your friend who watched over you overnight left your villa to fetch some breakfast, you intuitively knew that she was still there somewhere, outside, even when you could not see her any more, didn't you? You did not start imagining that she had turned into smoke and travelled back in time to hunt a dinosaur and cook one of its legs before coming back for you to eat it. That would be rather cool, I agree, but just as it would be unwise to jump off a cliff, or from a window, it is not going to happen. A fundamental reason *why* this won't ever happen is very tricky to state and prove, but we have to assume a few things if we are to try and unravel the mysteries of our universe. So, the first assumption, or "postulate," we will make is this: that we are able, somehow, to understand nature, even beyond what our senses can tell us. In order to do so, we shall henceforth assume that, under similar conditions, nature obeys the same laws everywhere in space and time, whether here or out there, whether now or in the past or in the future, whether we can see it or not, whether we know these laws or not. We shall call this our **first cosmological principle**. It is in bold letters because it is important. Were we not to assume it, we'd be completely stuck, and unable to guess anything about what happens in places where we are not looking, or which are too far away from us, or too far back in time. If we don't make this assumption, then your friend might as well have been time-travelling to hunt a yummy dinosaur.

In fact, there are many hints that this first postulate is correct, at least within the universe we see through our telescopes.

Take the Sun.

We know what particles, what frequencies of light, what types of energy come out of it. We detect them as they shoot out from its surface and land on Earth. So what about other, faraway stars? Do they shine thanks to the same kind of nuclear-fusion reaction or are they completely different? Are they like a burning log surrounded by fire, or are they made of plasma, like the Sun? We don't have many tools at our disposal to probe such queries. In fact, we only really have one: the light we receive from these stars. In it is encrypted many of their secrets, and one of these secrets that we have been able to decipher is that the laws of physics are the same everywhere. So, since light is key to our understanding of the cosmos, let's have a look at what it is.

Light, a.k.a. electromagnetic radiation, can be thought of both as a particle (a *photon*) and as a wave. As you will see later, both descriptions not only work, they must both be taken into account if we wish to understand our world. For now, however, it is sufficient to just look at it as a wave.

To describe waves in the ocean, you need to specify two things: their height and the distance between two consecutive crests. That height matters is obvious: you would be wise not to react in the same manner to, say, an approaching 100-foot-high wave as to one that is just 1 inch high. The idea is the same with light, and the height of a wave of light is related to what we call its *intensity*.

Similarly, there is a difference between waves at sea that are hundreds of yards apart and waves that are very close together. That distance is called, rather appropriately, the *wavelength*. The longer the wavelength, the fewer the waves that arrive during a given time period, a number that is related to the wave's *frequency*. To intuitively feel that the shorter the wavelength (or the higher the frequency), the higher the energy involved, you can picture yourself in front of a dam: whereas a 10-foot-high wave hitting the dam once a month wouldn't be cause for you to worry, a similar wave hitting it ten times per second would. For light, it is the same: the shorter the wavelength (or the higher the frequency), the greater the energy carried by its wave.

Now, contrary to what our ancestors thought, our eyes are light receivers, not light sources. And they are not built to detect all the types of light that exist, neither in intensity, nor in wavelength. Too powerful a source purely and simply destroys your retina, blinding you in seconds. That is what happens if you stare at the Sun, or at some lasers, or any source of light that is too intense. We can only see waves of light that are not too intense nor too dim.

The wavelength-related limitation of our eyes is more subtle. Over the millennia during which our ancestors (and we here include those that existed long before they had a human shape) evolved, their light-detecting organs adapted to see what they needed most to survive. To pick a fruit, or to be aware of the presence of a sabre-toothed tiger, it was rather more useful to see the colours green, red or yellow than the X-rays emitted by falling stars near distant black holes. In short, our eyes have thus adapted to the light that is most necessary in everyday life. We would have become extinct long ago had we only been able to detect X-rays.

In the end, then, what our eyes can see today is rather limited compared to all the existing natural types of light. But the universe doesn't care. It is filled with all of them. Appropriately again, we have called *visible lights* those lights we can see, and we've given individual groups of them some further names: colours. The distinction between one colour and another can sometimes seem rather arbitrary, but a very precise mathematical definition exists, a definition based on a distance, their wavelength.

It is true that some animals' eyes have evolved differently, that some can see lights that are slightly outside what we humans can detect. Snakes, for instance, have infrared vision and some birds can detect ultraviolet lights, both beyond our human visual capabilities.* But no animal has ever built apparatuses to detect them all. Except us. And we've become quite good at it.

From the least to the most energetic, the lights that surround us are: radio waves, microwaves, infrared light, visible light, ultraviolet light, X-rays and gamma rays. Radio waves have very long

---

* In fact, recent research seems to show that our eyes do actually perceive some – normally invisible – infrared light. What our brains make of it, however, is unclear . . .

wavelengths, 1 yard to 60,000 miles or more between each wave, whereas for gamma rays the wavelength is shorter than a billionth of a millimetre – but they are all light. And all the telescopes we have ever built have been engineered to harvest them, wherever they come from, whatever their intensity, to enable us to look at the universe through all the different windows our technologies allow. When you stare at the sky, whether it is with the naked eye or through some telescope, you therefore capture and process waves of light that have been emitted in outer space, somewhere, by a faraway source. As I said earlier, it was through a 3D reconstruction of all these images, recorded and reported, that you travelled in Part One. But what you may not have noticed back then is that, although it certainly was a journey through space, it was also a journey through the past, because light does not travel instantaneously.

Now that is an interesting, if rather gloomy, question that your tropical-island friends might have asked you: haven't we all heard someone, somewhere, at a dinner party or elsewhere, trot out the line that the stars we see in the sky are actually all dead?

Is it true? Are all the stars dead?

Well no, they're not. Not all of them, at least.

Let's see.

Let's assume that a great-aunt of yours, a distant relative who loves to give ugly crystal vases to everyone at Christmas, lives in Sydney, Australia. Being slightly old-fashioned, she never sends any news to anyone, except for on her birthday, in January, when she mails everyone a picture of herself standing next to the mailbox she is about to slide the picture into. On the back of the postcard, she always writes:

It is my birthday today.
Would love to hear your voice.
Love from your Auntie.
PS: I hope you liked the vase I sent you.

The problem is that even though you promise yourself, every year, to think about her, you don't, and, as always, by the time you receive the postcard it is not "today" for her any more. It may not even still be January. As usual, you hope she hasn't been sitting by the phone all that time, waiting . . .

In any case, what is important in this story is that the picture she took of herself a minute before mailing the card, the picture you are now holding in your hands, is unlikely to still correspond to what she looks like *now*. She may even be dead, for all you know, like some of those stars up there in the sky. Now don't worry, your great-aunt is sound, you'll get a few more vases, and you'll be able to have a few more goes at getting her to use email instead of post-cards. That would certainly be quicker. But it wouldn't be *instantaneous*. Nothing is. With email, you'd still get her picture a fraction of a second after it was sent. So once again, she might be dead by the time you receive it.

The idea here is not to make you paranoid that everybody you know might be dead. Rather, it is to show what happens in space, where the quickest delivery service possible uses *light* as a commu-nication device. And light, however fast, is very far from travelling instantaneously. In outer space, its unrivalled speed reaches a stag-gering 186,282.397 miles *per second*. Light can travel about twenty-six times around the Earth while you read this sentence. It is fast, the fastest thing there is, but amazingly slow considering the intergalactic distances involved out there.

Anytime a star shines, its light carries an image of itself. That image shoots through space at the speed of light, and it can take a very long time to reach us. This means that yes, the furthest stars we see in the sky are probably dead. But not all the stars. The Sun, for one, isn't. To be more precise, right now no one knows, but the Sun wasn't dead eight minutes and twenty seconds ago.

As you saw in Part One, the light from the Sun takes about eight minutes and twenty seconds to travel the 93 million miles that separate us from it. This means that if the Sun was to stop shining *now*, we'd know about that (rather big) problem in eight minutes and twenty seconds. It also means that, from the Earth, you will always see the Sun as it was eight minutes and twenty seconds ago. Never as it is *now*. The Sun that shines in the sky on a sunny day never really is as you see it *when* you see it. It isn't even *where* you see it any more. During the eight minutes and twenty seconds its light takes to reach your skin, the Sun moves by about 73,000 miles on its own orbit around the centre of our galaxy.

Now, the furthest light that we have managed to detect in our universe has travelled for as much as 13.8 billion years before hitting our telescopes, straight from when the universe became transparent.

The huge stars that started shining a few hundred million years after that most certainly do not exist any more, even though their light reaches us now, making them visible to us.

The same can be said for many other stars between the Sun and those far reaches of our universe.

On January 24, 2014, for instance, astronomers saw a star explode in the night sky, in a faraway galaxy. They saw it live, as the light from the explosion reached their telescope. As far as we're concerned, that star therefore died on January 24, 2014. But anyone living next to it would have witnessed the explosion as it unfolded there: 12 million years ago.

•

No one can travel to the other side of the universe. No one can teleport there instantaneously. In the end, probing the night sky is like receiving individual picture postcards from everywhere, all stamped at various times and locations in the past history of our universe, according to when and where they started their journey. Only by patching together all these postcards from the edge of forever can we reconstruct a slice of the history of this universe of ours, as it is seen from the Earth.

It is through such a slice that you travelled in Part One.

Until September 2015, to collect information about outer space, our technology did not leave us much choice: We had to use light. We had no alternative way of probing the far reaches of the cosmos. But this has now changed. A new tool has successfully detected a signal that had remained elusive until now. A signal that is not carried by light. As was announced on February 11, 2016, ripples of the very fabric of our universe have been detected, measured, and analyzed. These ripples are not made of light. As you will soon see, they are made of space and time, which they stretch and compress as they flow through everything at the speed of light. These very special wave detectors are a new window for us to probe our reality through: We are now able to detect what cannot be seen using light. And if you wonder what that might be, well, black holes and the Big Bang are good guesses.

We do not yet know what our new eye will reveal, though. So, before you get to learn a bit more about these waves and their monstrously powerful sources, let's see what we have already understood by capturing the lights that reach us from outer space.

To repeat: as of today, what we know about the history of our universe comes from the light that reaches us.

To decipher it, to understand it, we therefore need to figure out exactly what information light carries, and how it interacts with the matter and its building blocks – the atoms – that it meets in space.

You will plunge right into the heart of atoms in the part after next of this book but, for now, you do not need to know everything about them. Let's just say that atoms can be described as round nuclei surrounded by swirling electrons, and that these electrons aren't randomly scattered, but rather organized in layers around the nucleus.

It might be tempting to picture them as planets whirling around a central star, but that would be misleading – in fact we call the trajectories of electrons around their atomic core *orbitals* expressly to distinguish them from planetary orbits.

Given the right speed, a planet can theoretically orbit its star at any distance it likes – but this is most certainly not the case with electrons. Contrary to planetary orbits, electronic orbitals are separated by electronic no-go zones, places where electrons simply can't be. Moreover, the electrons can also easily – spontaneously, even – jump over these forbidden areas, from one orbital to another.

However, and here is the key point: such jumps do not happen for free.

To move from one orbital to another, electrons have to either absorb or emit some energy.

And since it so happens that the further away from an atom's nucleus an electron is, the greater the energy it carries, for an electron to jump from one orbital to another further away, it has to gain some energy, a bit like a blast of flame gives a hot-air balloon a lift.

Conversely, to move closer to a nucleus, an electron has to emit something, to get rid of some of its energy – like a balloon venting hot air in order to drift back to Earth.

But where does this energy come from?

Well, that is where light comes in: electrons can jump from one orbital to another by absorbing or emitting some light. But not just *any* light.

To get from one orbital to another requires electrons to jump over the electronic no-go zones that separate them, and achieving such a feat involves swallowing, or giving away, a specific amount of energy, corresponding to a specific light ray. Were they to be hit by some light that is not quite energetic enough, the electrons wouldn't be able to make the jump and they'd stay where they are. Conversely, hit by light rays that are *too* energetic, they could jump over several such zones and even be expelled from the atom they belong to.

This was figured out by mankind at the beginning of the twentieth century.

It may not seem to be groundbreaking but it is.

Einstein (he truly was everywhere) received the 1921 Nobel Prize in Physics for figuring this out as regards the atoms that make up various metals.*

•

Decades of experiments (and thinking) done ever since on all the

---

* Metals emit electrons *only* when they are illuminated with the "right" light. This is called the *photoelectric effect*. The explanation involves what I just described to you (electrons can only move from one orbital to another by climbing – up or down – energy levels) and the fact that light can be described as little packets of energy, like a particle. You will hear a lot more about this aspect of light later on in the book. And while we're at it, let me just add that Einstein probably would have deserved at least two other Nobel Prizes, but only got this one.

known atoms in the universe then made scientists realize that the energy needed for any electron to move from one orbital to another within any kind of atom is specific to the atom it belongs to. And this is very, very fortunate for us, because different energies correspond to different light sources – and with our telescopes, we can of course collect light from pretty much anywhere.

This simple fact means that scientists can tell what faraway objects like stars or gas clouds, or even distant planets' atmospheres, are made of, without ever going there.

Here is how.

Imagine a perfect source of light, one that shines all possible wavelengths of light, from the least energetic (microwaves) to the most (gamma rays), in all directions. Such a perfect source creates a glowing sphere of brightness. If there is an atom standing some distance away, its electrons, blinded by all the incoming light, may, in a frenzy, swallow all the ones they need to jump from where they are to a more energetic orbital. When they do so, they get excited.

Excited?

Yes. *Excited*. That is the proper technical term for what happens.

They are a bit like children being offered sweets at a party. And just as it is not hard to work out which sweets the children preferred afterwards (you only need to check what is left), you can figure out which types of light an atom gobbled up by checking which ones are missing in its shadow. All the unused light passes through the atom unhindered and you can detect their signature wavelengths quite easily. The ones that are missing, on the other hand, appear as small dark patches on an otherwise continuous rainbow of colours and light. Such a chart is called a *spectrum,*[*] and the dark patches are called *absorption lines.*

Scientists can tell, just by looking at the light wavelengths that are missing from a spectrum, which atoms are standing in the way of a light source.

Using light, you thus have a way to figure out what type of matter is out there, without going there.

And all the light-gathering telescopes mankind has used so far tell us that all the stars in the universe are made of the same stuff as the Sun, and as the Earth, and as ourselves. All the cosmic objects of the night sky are made out of the same atoms as we are.

If it were not the case, our telescopes would tell us so.

The laws governing nature can thus be presumed to be the same everywhere.

This is why the first cosmological principle is considered correct by everyone.

What a relief!

In fact it is such a good piece of news that being in outer space, you decide to immediately have another look at faraway galaxies, to figure out for yourself what they are made of. Aren't they pretty, with their beautiful spectra full of lines corresponding to hydrogen, helium and . . .

Now wait.

Hold on.

Something is wrong . . .

Looking at the spectra you gathered, you realize that the missing lines in the light coming from faraway stars are there all right, but they are not *where* they should be . . .

Whereas the electrons of some chemical elements here on Earth

---

[*]   To be precise, this is an *absorption* spectrum. A spectrum that shows what light a material emits, rather than absorbs (as is the case for our atom here), is called an *emission* spectrum.

are excited by blue light, the same electrons within the same chemical elements out there, in faraway galaxies, seem to relish slightly greener hues to jump from one orbital to another . . .

And atoms that are yellow-hungry here on Earth seem to prefer orangeish light everywhere else.

And orange-craving ones here like it red out there.

Why? How can that be?

Are all the colours shifted in outer space?

Or did we make a mistake?

You look again at different faraway sources. But there is no doubt. All the colours are shifted towards the colour red.

And it gets even worse: the further away the source of light, the more pronounced the shift . . .

Damn. It was all too easy.

So what is going on?

Are the laws of nature different in different areas of the universe after all? If you could stroll the land of a planet similar to Earth, a planet orbiting a sunlike star billions of light years away, would its sky and oceans and sapphires be green, would the plants and emeralds be yellow and its lemons red?

Well, no.

If you travelled there, you'd see the alien world as we see it here, a world in which lemons are yellow and skies blue. The reason for the observed colour shift is not that the laws of nature are different far away from us. It goes deeper than that. It even changed everything mankind had believed for more than 2,000 years.

Have you ever tuned a guitar, or any other stringed instrument? Did you notice that the note emitted by a plucked string changes when one adjusts its tuning pegs? The more one stretches the string, the higher the pitch, right?

Well, what you just saw in the sky corresponds to the same phenomenon, except that sound is replaced by light, and the string is not a string. In space, light does not travel, or *propagate*, on a string, but through the fabric of our universe itself. And to explain the colour shift you just detected, this fabric has to be involved.

Why?

Because for this shift to affect all the possible colours in the exact same way, it cannot be the case that light itself is to blame, but rather what it travels through.

Pluck a string and tighten it with the tuning knob, the sound it emits is shifted towards a "higher pitch," not because something happened to the sound, but because the string got stretched. And a guitar string is stretched in the exact same way for all the notes.

Now imagine you could stretch the fabric of our universe, like you would stretch a guitar string. Stretch it once and all the wavelengths of all light propagating on it would immediately become "higher-pitched." Why? Because light can be thought of as a wave and the stretching would increase the distance between two consecutive crests, the wavelength. Blue would become green. Green would become yellow, yellow red and so on.

On a spectrum, it means that the actual colours of the universe get shifted towards the colour red. They are *redshifted*.

Now instead of stretching the fabric of our universe once, imagine it somehow has been continuously and steadily stretched. The further any light has been travelling, the greater the redshift it will have been subjected to before reaching Earth. With such a scenario, starting far away, a blue ray will steadily become green, and then yellow and then red and then invisible to our eyes: infrared and then microwave . . . Knowing how much the colours emitted by a faraway star differ from their initial colours once they reach Earth would then enable you to tell how far away that star is.

But is this true? Is that how the fabric of the universe behaves?

It is. That is exactly what you saw in the sky.

But what does that mean in practice?

It means that the actual distance between faraway galaxies and us is growing all the time. It means that space stretches, and therefore grows, on its own, in between the galaxies. It means that our universe changes with time.

Countless experiments have now confirmed this, and scientists have learnt to accept that idea. We *do* live in a changing, growing universe.

Einstein didn't like it though. Nobody liked the idea, a century ago. For our ancestors, be they scientists or not, the universe had always been the same. But they were wrong about that.

To be clear, it is not the galaxies that are moving away. It is the *distance* that separates us from the already faraway galaxies that is growing. It is the very emptiness of space that is being stretched. Scientists have given that phenomenon a name. They call it the *expansion of the universe*. And contrary to what one might think, this does not mean the universe is expanding into "something." It means it expands and grows from within.

Now, before drawing any hasty conclusions and wondering what could have caused such an expansion, you may want to check all this by yourself. So imagine you are rich as can be (100 billion dollars in the bank, say) and that you have a hundred friends. Being curious about our universe, you give each of them a billion dollars to buy a powerful modern telescope and travel all around the Earth to gather the light from as many faraway galaxies as they can.

A few months later, you invite them all to your mansion to present their finds. About half of them were real friends and show up (you can consider yourself rather lucky), the other half having preferred to keep the money. But it doesn't matter, for all their stories are identical. Wherever they went, whether in China, in Australia, in Europe, in the middle of the Pacific or in Antarctica, all those who came back to you saw the same phenomenon in the sky: right above their heads, faraway galaxies had a strange colour-shift. They were all receding into the distance. And the further away these galaxies were, the faster they were fleeing away. They all witnessed the expansion of the universe.

What should we make of this?

As you think about it, the peculiar feeling you had at the end of the last part again grips your mind.

First there was this strange visible universe that was a sphere centred on you, and now this . . .

Could it be true?

If everything, everywhere, moves away from the Earth, does it mean that all the mothers on Earth are right in thinking that their child is at the centre of the universe?

As amazing as it sounds, it does seem so.

What a lovely piece of news, what a joyous day.

If some of your friends are around as you read this, you can uncork some Champagne. We *are* special, after all. Especially you.

At last. Vindicated. Copernicus was wrong. He should have listened to his mother. Mothers are always right. We all are at the centre of our universe.

But wait, wait, wait . . .

How about mothers on faraway planets, in other galaxies?

Were they to exist and think like our mothers, would they be wrong about their children?

Or is that a proof that there aren't any mothers anywhere else? Surely not.

Despite what you have seen, just as Copernicus told us 400 years ago that we are *not* at the centre of the solar system, most (if not all) scientists today take it that our position in the universe is no more significant than any other. Strangely enough, this doesn't mean we're not at the centre of our visible universe. We are. But every other place also is. Every place is at the centre of the universe that is visible from there.

This very strong belief even led scientists to the following additional cosmological principles:* to guess what happens out there, very, very far away from our planet, scientists assume that there is no preferred position anywhere whatsoever – that is the **second cosmological principle** – and that if one selected observer was to travel around, all directions would always look the same for him or her, with faraway galaxies always moving away from where he or she is, just like they are moving away from us here on Earth – that is the **third cosmological principle**.

If you now take a moment to think about it before your friends give up on the Champagne, cosmological rule number three sounds trivially wrong.

The world obviously doesn't look the same as seen from where you are now, reading this book, as from under your shower (assuming you are not reading this while taking a shower). So a clarification is in order: the third cosmological principle does not bother about what is near you. It is concerned only with the big picture. With scales much, much bigger than galaxies. It says that the universe, on extremely large scales, looks similar whatever direction you look in.

Still, this sounds wrong, doesn't it? Didn't you travel throughout the universe in Part One? Didn't you see places faraway that did not look like the universe as seen from the Earth? You even crossed a patch of space thousands of light years thick where no star shone, the so-called cosmic Dark Ages. How can the universe look the

---

* Remember, the first cosmological principle was that the laws of nature – whatever they may be – are the same everywhere.

same from the Earth and from a place where there are no stars at all?

Well, it is now time for you to realize what I truly mean when I say that you did not travel, in Part One, through the universe as it is, but through the universe *as it is seen from Earth*. It is not quite the same thing. Remember: the universe that appears at night does not correspond to what our universe is *now*. It corresponds to a slice of its past history, a history centred on Earth because we are on Earth. We receive picture postcards every day, from everywhere. According to cosmic rule number three, aliens living on a faraway world should see a universe exactly similar to ours. Not in detail, of course, but on large scales. They, too, would be surrounded by the sum of all the information that reaches them from their past, they too would see in their night sky a slice of the history of our common universe. They'd have their own cosmic Dark Ages and surface of last scattering. They'd have it all, even if their slice did not intersect our own.

In the end, to understand our universe, to get the whole picture, all the past histories of all the points of the universe have to be added. Close-by places have histories that overlap a great deal, of course, but places separated by great spatial distances may not have any of their pasts overlap at all. Still, they should all be considered equivalent. That is what cosmic rule number three means in practice. You will hear more about this later.

Incidentally, this also means that even though you do not occupy a special position in this universe of yours, you still are – as your mother certainly thought – at the centre of *your* visible universe.

And if you feel like you had always known so, please let the joy flow through your body and mind. It is great news.

I say it again: you *are* at the centre of your universe.

Now, what may feel less good is that so is your neighbour: he or she is at the centre of his or her visible universe.

And everyone else too.

~~And everything else as well.~~

We all are, everything is, at the centre of our own universe, the universe we can probe with the light that reaches us. Only on some rather special occasions can two people's visible universes perfectly match. I will leave it to you to figure out when, and how, that can happen.

So, that being said, it is time to look a little closer at this expansion that stretches the universe.

Is that really what is happening?

It is. The distances between faraway galaxies do stretch all the time. It doesn't apply to close-by objects, though, because gravity is locally stronger. Galaxies create a gravitational attraction that cancels out such expansion, both within their boundaries (the distance between the Sun and nearby stars is not expanding) and around (neighbouring galaxies are actually getting closer and closer, all the time). Over large distances, however, expansion rules.

The discovery of the expansion of the universe was made by US astronomer Edwin Hubble, in 1929, and the law that links the way the galaxies recede to their distance away from us is called *Hubble's law*. On the strength of this finding, Hubble can rightly be considered one of the fathers of modern observational cosmology. He is also the person who, with Ernst Öpik, proved that the Milky Way is not the entire universe, that other galaxies exist beyond it. Two discoveries that would certainly be worth a Nobel Prize, were they to be made today. At the time, however, looking at the stars and trying to make sense of them was not considered part of physics, either by the physics community or by the Nobel committee. As a consequence, Hubble never got a Nobel. But the rule changed after his death, and many a Nobel Prize has since been awarded to observational cosmologists. You shall meet some of them in this book.

Now, as you are about to understand an extraordinary consequence of Hubble's expansion law, you will most probably be shocked by how bright scientists can sometimes be. With a lot of thinking and about twice as much coffee, they figured out that if everything that is far away in our universe is moving away from us now, then everything that is now far away must have been closer in the past.

Wow.

Talk about breakthroughs.

You may try and do that reasoning again on your own someday, it is quite satisfying.

Actually, although it may not seem to be much, it was a real revelation.

As I said above, Einstein himself refused to believe it.

Why?

Why does it matter if faraway galaxies are moving away or, for that matter, were closer in the past?

Remember: Hubble's observation-based law says that it is the distance between the galaxies itself that is expanding, not just that the galaxies are moving away from one another.

In other words, it is the fabric of the universe that is expanding.

Following this idea through, it must be the case that the universe in its entirety was smaller in the past.

But how could that be?

And can one prove it?

One can. By looking far away again. The past lies there for us to receive its messages. And the wall you saw at the end of the visible universe confirms all this brilliantly (although it is dark), and you will see why in the chapter after next. First, however, you will have to travel to outer space again, to get a bit more acquainted with gravity.

Of the four fundamental forces that govern our universe, gravity is perhaps the one we are most aware of.* Every time you fall, every time you use the muscles in your legs to push yourself up to standing, every time you lift anything, your body is reminded of gravity's existence.

Everything is affected by gravity.

But everything also *creates* gravity. Including you, including those crystal vases your great-aunt in Sydney keeps giving you at Christmas.

Talking of which, imagine you had one of her vases with you on your island.

Look at it.

Now drop it above a hard surface.

It falls and shatters to pieces.

You can imagine letting all your collection fall onto a hard surface, from as many different places on Earth as you can think of.

Amazingly, they would always fall. And break. Wherever you are. Good.

Not only will such an experiment rid you of your vases, it'll also prove a point: as long as it is denser than air, any object dropped on Earth will fall, just as Newton (and everyone sane) has thought since for ever.

What about an object lighter† than air, then? Why do helium balloons rise in the sky rather than fall? Do they not feel Earth's gravity?

They do. But there is competition.

_____

* You will learn about the other three forces very soon, starting in Part Three.

† "Lighter," in this chapter, is to be understood as "less dense."

Whenever objects are dragged down by the Earth, the densest tends to settle furthest down. If objects lighter than air seem to fly up, it is because the air from above is denser and takes their place. Were the air to be visible, you'd see it. But visible it is not and you just see the result: objects lighter than air are pushed upwards, by the invisible air that stacks up under them. Gravity is always attractive. It always makes things fall. But competition creates layers, and some objects have to move up to make room for denser ones.

With this in mind, you can think of the Earth as a huge ball with loads of stuff sticking to its surface because of the steep curve it creates around itself within the fabric of our universe. All the objects you've ever seen slide down that slope, *you* slide down that slope, until a floor or something else denser prevents them and you and everyone from sliding down any more. Rocks in the Earth's crust are denser than water. That's why the ocean lies on top of the hard rock beneath. Rocks and water are denser than air. That's why the atmosphere lies on top of our planet's surface, be it rocky or liquid.

We humans live underneath about 62 miles of air sticking to our planet's surface. We are denser than the air. We don't fly. But we are lighter than the ground. So we stay on top of it. Sometimes, some objects or animals do manage to move away from the ground, up in the sky, but for them to do so requires energy and it usually doesn't take long before they fall back, unless, of course, they are lighter than air, which is unheard-of (and would be rather unfortunate) for any animal.

Now, how would everything fall into place if there was no Earth?

It is Sunday morning on your tropical island. Your friends have been bringing you breakfast every morning since your strange mind trip, and they have clearly been more and more curious about your story. Some of them are even wondering whether you really did see what you keep saying you saw. Others have been struggling to get to sleep at night, worrying about the death of the Sun. Rather unfortunately, these ones have been actively looking for ways to stop you talking about it all the time. And it seems they did find one.

You open your eyes.

~~Motes of dust flicker and dance in the morning sunrays even though~~ they, too, feel gravity, you think, as someone knocks on your door.

"Come in," you say, sitting up on your bed, expecting a smiling friend and perhaps a tray of fruit and coffee.

The door opens. And there she is. Your great-aunt. From Sydney.

Next to her are three bags, all filled with crystal vases. You did not think it possible, but they are even uglier than the ones you wanted to smash for your gravity experiment.

She comes in, not in the least bit fazed to find you in bed, stands next to you, pats you on the cheek and hands you one of the vases, silently smiling, with a look of understanding on her face, knowing that words could not faithfully convey your joy at her surprise visit.

With the vase in your hands, you close your eyes to keep calm, suddenly desperately wishing you were somewhere else.

And as you open your eyes again, you are.

Somewhere very else.

In outer space.

The villa, the sunrays, your bed, your great-aunt, all are gone.

You are back among the stars, as in Part One, but everything seems much safer than back then.

You can't help but break out into a big smile as you look around.

No sign of an immediate explosion.

No molten Earth.

All the stars are far away, all is quiet.

You are floating in the middle of a seemingly infinite darkness studded with tiny lights.

When, in the first part of this book, you found yourself in space, you were just a mind. Apart from the moment you got ejected by a black hole, you did not feel anything at all. This time, however, you are about to experience something different. You are still on a mind trip of sorts, but you did not leave your body behind. It is here, wrapped within the protective shelter of a spacesuit's fabric, experiencing weightlessness.

It all seems so real that you actually feel rather sick, but you soon get over it, and it eventually occurs to you that although your great-aunt is no longer around, you're still holding the vase she just gave you.

You look around again with a grin, but there's nothing at all to smash it against. No Earth. No star.

Putting on a brave face anyway, you decide to do yet another gravity experiment.

You open your hand, at arm's length, and let the vase go. As far as you can tell, the vase stays right where it is. A minute passes. And then another. And then, suddenly, after another minute goes by, another minute has passed with still nothing happening.

Or maybe the vase moved a bit towards you. But not by much. Nothing worth reporting.

Eventually, tired of staring at this monstrosity of a vase, you push it with the tip of your finger and watch it as it slowly moves away in what appears to be a straight line. Good riddance.

Had you not pushed it, the vase would have stayed right next to you. It wouldn't have fallen. What could it have fallen towards anyway? With no planet or star around, there's no notion of up and down, or right or left. In the middle of nothing, all directions are equivalent. There's no ground of any sort for the vase to head to, unless, of course, you consider *yourself* to be the ground. But that would be to insult yourself, wouldn't it? Well . . . you shouldn't take anything too personally when nature is involved, for after a long while spent doing nothing, to your great dismay, you do see the vase coming back towards you. Gravity is at work. The gravity *you* create.

A strange question pops into your mind, though: is it the vase that is moving towards *you* or *you* towards *it*? For all you can tell, it may as well be the vase that is the ground and you who are falling towards it. Unfortunately, you do not have the time to dig that idea through because an asteroid shoots past right next to you, snatching you and the now rather close-by vase with its invisible gravitational fingers.

Had you been asked, you'd probably have said that, being heavier, you'd be first to hit the asteroid's ground. But no. That is not what happens. You and your vase reach that rock's dusty surface simultaneously and, as your feet hit the soft ground, you immediately grab that failed piece of art, to smash it against the asteroid's surface.

Unfortunately, the asteroid's ground isn't as solid as Earth's, and the vase doesn't break. Instead, a vast cloud of cosmic dust now surrounds you . . . Annoyed, you pick the vase up and throw it spaceward, with all your strength, to get rid of it once and for all. This time, there's no way it'll come back, you think, and you do feel relieved as it disappears into the distance, through the dust cloud, doomed to spin on itself forever.

Alone at last!

You can now relax, enjoy the unspoiled view and figure out how to experience gravity more deeply than anyone has before.

As you wonder about this, it occurs to you that the rock you happen to be standing on is not moving straight any longer. Its trajectory just bent to head towards a dark and frozen world, a planet without a star, which wanders in the middle of nowhere on a probably vain quest to find a shining new home. There was danger around, after all. You just hadn't spotted it.

For a moment, as your rock accelerates down towards the planet, as you feel your guts lurch up inside you, you are almost certain that you are on a perfect collision course, on your way to smash into the cold and very dead world's surface. You've heard that when facing imminent death, people usually have long-lost memories return, or see their life unfold before them. No such thing happens to you though. You can't think of anything but your great-aunt's face as you blame her and her vase for the certain death that awaits this body of yours.

In a heroic effort to save your life, you push down hard and jump from the asteroid, and start swimming away from the planet. Right after doing so, you realize two things: contrary to what you thought, you are not on a collision course, and secondly, although jumping from an asteroid is certainly possible, it is impossible to swim in space.

As if on an interstellar rollercoaster ride, you accelerate more and more as you slide down the slope the planet creates in the fabric of the universe. As expected, you end up missing its surface by a few thousand miles and swing around its dark, cold ground only to find yourself rapidly hurled spaceward again with your asteroid, like a slingshot, at a much greater speed than before the fall. You and your asteroid effectively just stole some energy off that world, some kinetic energy, like a golf ball on a mini-golf course that, missing a cunningly moving hole, spins round its edge before being ejected and rolling *faster* and discouragingly *further away* than where you hit it from. A still hole cannot do that, nor can a fixed world. But a moving hole can, and so can a moving planet.

A few minutes later, as the dead planet disappears into the distance, you land back on the surface of your asteroid. Strangely, you realize, it had never stopped attracting you, and, stranger still, you notice that you both followed a very similar path around the now-gone lost world.

That a vase weighing a fortieth of your weight should fall like you towards an asteroid may be surprising, but that an asteroid, a rock the size of a small mountain, should fall like you towards a planet is vexing. Still, that is what happened. Objects, it seems, all fall in the exact same way towards planets, or towards each other, regardless of their mass. As curious as it may sound, even the Sun and a feather would fall in the exact same manner towards an asteroid, or a planet, or anything. It is so, because being subject to gravity means journeying down the slopes that matter and energy create in the fabric of our universe.

Understandably, you sit down on the rock to let this thought settle into something meaningful.

You stare at outer space.

No meaningful idea comes to mind.

You keep trying and, tenacity eventually paying off, an extraordinarily beautiful picture suddenly emerges in your mind.

You start to see curves and slopes and hills everywhere, around rocks and faraway planets and stars and galaxies. Light rays coming from bright faraway sources seem to slide around these slopes, leaving evanescent fluorescent lines on their paths, for you to picture them, for you to see the real shape of the universe's canvas. You see that, just like matter, just like you, light in space does not travel along straight lines, as you might have thought. Near a galaxy, a star, a planet or even a little rock, light gets deflected. The denser an object and the closer a ray swooshes by it, the more pronounced the deflection. As the planets and stars and galaxies move, so too do the curves and slopes that they create, following them as they dance around one another and merge. Everything moves, everywhere, in this universe of ours. Even its fabric.

It seems to you that this very fabric whose shape you are seeing, this fabric which has remained invisible to you up to now, in fact looks almost alive.

Watching all this as you sit on your asteroid, you are sliding down a curve, just as you are right now as you read this book. On the asteroid, it is the rock that creates it. While reading this book, it is Earth. On the asteroid, the curve is gentle and it would not require much energy for you to fly away from it. On Earth, the curve is steeper.

If you do not have the impression that you are falling as you read this book, it is because there is ground under your feet, or a chair you are sitting on, that prevents you from doing so. But you probably do feel that your shoulders (your whole body, really) tend to be dragged down. All the time. However, if you are reading this while freefalling from a plane, then you'd really be falling down the curve created by the Earth although the presence of air would slow down your fall. Such a fall down a slope in the fabric of the universe is the most natural motion of all, for and around all objects throughout the universe.

When you first pushed your ugly vase away, it slowly climbed the invisible slope *your* presence created, and then it fell down that very same slope, just like an object thrown upwards from the surface of Earth slows down as it moves up, and then speeds up as it falls down.

To make it to space from the surface of Earth, an object has to be thrown vertically at more than 25,000 miles per hour. Shoot it slower and it will fall back down.* Always.

To escape your gravitational attraction (not to be confused with your attractiveness), a minimum velocity is also required, just as a minimum initial speed is required if one wants to roll a child's marble up and over a bump in the ground.

You did not push your vase fast enough, so it came back to you, because you, too, curve the fabric of the universe.

And later, when you whizzed around the planet and swung out

---

* Bullets fired from any rifle are way slower than this, so they always fall back, even if you shoot towards the sky. So don't try it. This 25,000 mi/h velocity is called the Earth's *escape velocity*. By way of comparison, the Sun's escape velocity is about 1.3 million mi/h, whereas the rubber-duck-shaped comet upon which the European Space Agency space probe Philae landed in 2014 has an escape velocity of just 3.4 mi/h. A little jump would be enough to get away.

the other side with a little extra kick taken from the planet's own movement, you unknowingly used a technique that space-rocket scientists use to send satellites far away in the solar system without needing any fuel: by having their apparatus fly near to planets at the right angle and distance, they can get them slingshot towards deeper regions of our cosmic neighbourhood, with boosted velocity.

As these thoughts wash over your brain, you now understand that even on Earth, everything is indeed falling all the time down the slope created by the matter our planet is made of. That is how and why our planet is layered, from the top of its sky to its inner-most core, with the least dense particles above and the densest ones buried deep inside. It took billions of years for such an equilibrium to be reached.

Now, whether you are aware of it or not, you just got rid completely of the idea of gravity being a force. Rather, you now see it as a landscape of curves and hills and slopes, and it seems that this was the lesson you just travelled to space to learn, for as soon as you think this, you suddenly find yourself back in your villa, lying down on your bed, facing your great-aunt – who seems rather confused.

"Did I not just give you a vase?" she wonders, seeing none in your hands.

"What vase?"

"Never mind, dear, never mind."

That night, as you eventually manage to find some time to be alone, you escape the relative civilization of your villa to stroll the beach and face the stars. Your great-auntie's comment about gravity niggles you, and you try to summarize what you've just learnt.

There are slopes in the very fabric of the universe.

Everything creates a slope in every direction, an invisible slope that we call gravity, and the denser the object that creates it, the steeper the slope. But if all massive objects bend the fabric of our universe, then surely light also does, you think, because energy is mass and mass is energy, as per $E = mc^2$.

But is that really true?

Does everything really bend that fabric, light included? And what the dickens could that fabric be made of?

Back there, in your villa, or anywhere you've ever been, have you ever felt it? Have you ever felt the invisible slope created by a wall? By a sofa? By a ceiling? Or by the sky? Or by the light coming from a lamp? No, you have not. You've only ever felt the one created by our planet as a whole, the one that your muscles and bones fight in order to get up in the morning. Were you made of pure water, you'd splash and spread on the floor, not on the wall.

In truth, whatever gravity you are feeling right now is actually the sum of *all* the slopes created by everything that surrounds you, including the walls and the ceiling and even a bird or a plane that might fly high above your head.

But everything that is below you right now is far more significant than anything that is above. The Earth beneath your feet contains more matter and stored energy than the sky above your head. So it creates the steeper slope. Hence you are inclined to slide down that one first, and feel it most. That's Earth's gravity.

But what about the fabric of the universe? What *is* that fabric? *What* is bent?

Well, that is actually what Einstein figured out.

With $E = mc^2$, he showed that the distinction between mass and energy is superfluous, that mass and energy are but two aspects of the same thing. That was in 1905. In 1915 he showed that the shape of the universe in any given place is determined by the mass and energy that is present in that place. In passing, he got rid of the idea of gravity being a force. Gravity is just geometry. Curves and slopes. Created by matter and energy. But the geometry of what? There's no such thing as a cosmic stretching soapy rubber cloth upon which everything moves, that is clear, but remember: just because we do not see something, doesn't mean that it does not exist. Before mankind understood that the invisible air around us is made of atoms and molecules, everyone thought it was empty.

We here have the same kind of conceptual gap to bridge: outer space, despite its seeming emptiness, is not empty. Nor is it static.

What makes it a moving and changing geometrical object is precisely what I have until now called "the fabric of the universe."

Einstein discovered that this fabric is a mixture of space and time, two entities that, as we have learnt to accept during the past century, can't be separated.

The fabric of the universe is thus now better known under the name *spacetime*, and Einstein's general theory of relativity tells us how this spacetime is bent by what it contains, and vice versa. Energy and matter on the one hand, and the geometry of spacetime on the other, are identical concepts as far as gravity is concerned.

So far, however, you've only experienced the bending of space. Not of time. Or so you think. In fact, the bending of time was always occurring. It even takes place around you right now, as you read. Its effects are too weak to be noticed by your senses, but you will soon find yourself in places where the bending of time will be obvious, and very confusing. This will happen in an aeroplane, in Part Three, and when you eventually dive into a black hole, in Part Six.

For now, however, you are back on your beach and watching the stars. It is late but you do not care. You stare at the heavens and you certainly feel like you are floating in the middle of wonderful ideas that seem completely crazy, but which, for some pretty miraculous reason, seem to be very good at describing our cosmic reality.

Thanks to our planet's bending of spacetime, everything that is close enough to it falls down towards its surface and adds to the bending. Thanks to this, over the billions of years since Earth was born out of a cloud of stardust, an equilibrium has been reached where our planet got surrounded by an atmosphere, the one that now protects us from outer space, allows us to live by breathing air and gives us the opportunity, sometimes, to stare at the sky.

Just outside this atmosphere, away from Earth, there is our Moon, which spins around our planet like a marble spins in a salad bowl, with this difference: that the Moon itself also creates a bending in spacetime. The Moon's own bending of spacetime makes the water that lies here on the surface of Earth fall Moonward. That is why the water follows the Moon as it orbits our world, creating the tides.*

Further away still, there is the Sun, with the steep bend of its spacetime slope, down which all planets and comets and asteroids of the solar system are spinning and whizzing at different velocities and heights, like marbles on the wall of that salad bowl.

And then there is competition with our neighbouring stars.

Some distance away, other stars' bending of spacetime becomes more pronounced than that of our Sun, and those faraway comets that lie near the limit may sometimes reach the tip of the hill and move from the realm of one star to that of another, just as a marble thrown from the top of one salad bowl might fall down into another, if there is one nearby. In space, there is always one nearby.

And all the spacetime bending of all the stars in the Milky Way adds up to create our galaxy's bending, our galaxy's gravitational field, which competes with that of neighbouring galaxies, and then the Local Group itself competes with the aggregate bending of other groups, and so on. And Einstein figured out a way to understand this all in one single formula.

Well done, Einstein.

---

* The Moon also pulls everything else, of course, including our planet's solid crust, and ourselves, and teacups and spoons, but these are solids (and/or smaller) so it shows less.

His equation even made him forecast that strange, unheard-of waves should fill his immensity.

The first time you heard about gravity as being a bending was a few chapters back, when I told you that planets and stars were very much like heavy balls bending a stretched rubber band. But you now know that the fabric of our universe (the mixture of space and time we just called spacetime) is not a band, nor is it flat. It is everywhere. A planet or a star, in outer space, is therefore better represented by a ball not lying on a flat surface, but immersed in an ocean filling the whole universe. No surface above, no ground below. Just water, everywhere.

If such an immersed ball could *distort* the liquid in its vicinity, in all directions, dragging the water towards itself, then that would correspond to how gravity works. A fish peacefully swimming by would be pulled, together with the water, towards the ball. Incidentally, near the ball, the fish would not swim straight. It would be deflected. With the right velocity, it could even stop swimming and lazily settle into orbit around the ball. That is what happens in space: a planet requires no flapping of any fin to orbit its star. The Earth indeed moves straight, in a spacetime curved by the Sun. Our planet does not steer nor spend any energy to do so. It just follows the invisible curves of spacetime created by our star, like a marble in a salad bowl.

Taking this analogy one step further, you can ask yourself what would happen if not one, but two balls were immersed in the ocean, and orbiting each other.

They'd surely create some waves.

Not waves on the surface, but waves within the ocean itself. Those waves would then spread outward, away from the spinning balls, making the balls lose energy, until they collide.

Now, what would such waves correspond to in our universe? To a wiggling of its fabric. They would be waves of spacetime, and this is what we call *gravitational waves*. Their existence was predicted by

Einstein in 1916, just a few months after he published his theory of gravity. No one cared to listen to him, though. For decades. He eventually gave up on them, believing that they may well be an artifact of his calculations and not real, until French mathematician and physicist Yvonne Choquet-Bruhat told him, in 1952, that he was right. . . . If General Relativity is correct, she had mathematically proved, then gravitational waves must exist. The race to detect them was on.

Two black holes weighing 29 and 36 times the mass of our Sun (55 and 68 miles in radius)—1.3 billion years ago, in a galaxy 1.3 billion light years away—spiraled towards each other and merged, at half the speed of light. During the 20 milliseconds or so that the collision lasted, they lost the equivalent of 3 solar masses worth of energy. A colossal amount of energy. About 50 times the power of all the stars in the visible universe combined. According to Einstein's theory of general relativity and $E = mc^2$, this energy was turned not into light, but into gravitational waves that nothing could stop, destined to reach Earth 1.3 billion years later.

And they did. At 5:50:45 a.m. Eastern Time on September 14, 2015.

No one would have seen them had it not been for the extraordinary wit of physicists Kip Thorne, Rainer Weiss, and Ronald Drever, who spent decades of their lives to build LIGO, the US-based Laser Interferometer Gravitational-Wave Observatories that detected them. Following their lead, more than a thousand scientists from the world over took part in the search. No doubt a Nobel Prize will crown this extraordinary achievement in the years to come.

So yes, Einstein was right about this too. Quite a man. You almost wish he could appear now in front of you, so that you could shake his hand. And there still is more to his theory of general relativity. Did you not read, earlier, that Einstein opened the door to the idea that our universe might have a history? That our universe was smaller in the past?

You sit down on the beach and close your eyes, focused, ready to picture exactly what that might mean.

7 | *Cosmology*

There are some questions in life for which a single and uncontroversial answer can be given. Unfortunately, despite what you've just seen, what our universe looks like as a whole is not one of them. Einstein's equations allow for many different global shapes for our universe and, as you will see in Part Six, we do not even really know what our universe is made of.

Having said that, it may be noteworthy to remind ourselves that physics, however powerful it has been up to now, has never *exactly* matched reality. It even knows it can't aim to, because it would mean that reality – whatever it is – could also be *exactly* known. Which it can't. Observations and experimentations, whatever their accuracy, always give an approximate answer: there is always an error margin, however small that margin may be.

With hindsight, we even know that throughout human history, the technology with which we humans have probed nature has only rarely been in step with what physics could predict at the time, sometimes leading to wrong beliefs. If, a few hundred years ago, an ancestor of yours had somehow managed to guess the existence of some bacteria whose size are a thousandth the width of a hair, none of his contemporaries would have been able to check this out and he probably would have ended in an asylum, for unduly scaring people. The same goes for faraway galaxies. Had your ancestor also insisted upon their existence, he would not have been locked away but burnt alive. Like Giordano Bruno. The necessary technology to see far enough into the cosmos to picture them did not exist until less than a century ago. Likewise, the technology to check what you will see at the end of this book has not been engineered yet.

That being said, science progresses by steps, and sometimes by huge steps, opening the path for revolutions in understanding. Still, it may be healthy to look at science as being like scaffolding for thoughts, scaffolding that tries, generation after generation, to be as close as possible to the reality we live in, a reality whose mysteries are then unravelled through experiments. And it could well be worth mentioning that even though it may change in the future, so far no human activity but science has ever led to discoveries about nature that weren't there to be seen in the first place. However humble one needs to be before the majesty of nature, science, and only science, has given us eyes to see where our bodies are blind.

Contrary to what most people may believe, scientists don't like complexity. To try to understand the universe as a whole, they prefer everything to be simple. And the game, usually, is to figure out a simple pattern within a seemingly complex environment.

That's where some wit comes into play.

So let's see what we can make out of Einstein's vision by simplifying, on as large a scale as possible, everything you've seen so far. Let's forget about details. Let's have a look at the BIG picture. No asteroids, no planets, no stars, no gravitational waves.

They are far too small to be important for what matters here. Let only galaxies, clusters of galaxies even, remain. And you are there, able to see it all, as a longsighted eye of such cosmic proportion that the Earth, the Sun and the hundreds of billions of stars that make up the Milky Way are but a dot marking your position.

The other galaxies are evenly spread around you, even though structures that resemble filaments are apparent.

Good.

That is simple. This is your initial set-up. You feed it into Einstein's equations to see what, if anything, comes out of them. And you wait, anxious, not daring to expect much. And then . . . a miracle! It works! All around you, everywhere you look, the galaxies and clusters of galaxies move around each other as expected, but that is not all. The universe that surrounds you, the volume of the visible universe that can be observed from Earth, starts to expand. Spacetime stretches in between all the galactic dots, making them move away from each other independently of how they move around each other! Whatever their movement on a small, local scale, they are like poppy seeds in a cake being baked or dots on the surface of an inflating balloon: the further they started from Earth, the faster they move away. That is what your friends saw when you gave them their billion-dollar telescope. That's the expansion of our universe.

By feeding Einstein's equation with a simple model of the visible universe, you got something that, before Einstein's time, had never been imagined at any time in the history of mankind. Something that corresponds to what you saw up there in the sky, to what scientists see every day: the universe itself can (according to Einstein) and does (according to observations) evolve.

With such a thought, *cosmology*, the science of figuring out the past and future history of our universe, was born. Before Einstein, we only had *cosmogonies*, stories we told ourselves in order not to go mad about the mysterious origin of our reality. Now we have science too. A means to unravel the story that not humans, but nature, has written.

As you watch all the dots that surround you evolve, you suddenly realize that, with Einstein's equation, you can indeed press "rewind" in your mind to make the expansion roll backwards.

And you do.

Instead of inflating, the poppy-seed cake that is our visible universe immediately starts to deflate. Your cosmic eye sees it shrink: the pasts that were far away now move towards the present, towards you, swallowing away the images of the years to come.

It is the whole sphere limiting Earth's visible universe that shrinks.

And shrinks.

And shrinks, until . . .

About a hundred years ago, when Belgian physicist and Jesuit priest Georges Lemaître decided to implement the three cosmological principles into a similar, imaginary clockwork universe, in order to watch it expand and contract in time, his conclusion was straightforward: our reality, it seemed, the very reality that had been taken for granted since humans were able to think, probably had a beginning.

Einstein's equations had quickly led Lemaître, and many others afterwards, to the very confusing idea that our universe, even though it has always contained all the energy it still has today, once had no size at all.

No size in space or in time.

An idea that definitely sounded absurd, and arguably still does; but that was what Einstein's equation said.

From what we know today, however, this seems to be the best idea humankind has ever come up with to understand what we see in the night sky.

Any theory that asserts that everything our visible universe contains once had zero (or very close to zero) size at some stage in its past is called a (hot) *Big Bang theory*.

"Hot" because only a very hot past can cope with having all the energy of our visible universe squeezed into a very tiny volume. The heart of the Sun is hot because all the matter it contains is squashed by the Sun's own gravity. Shrink the whole visible universe into a Sun-sized sphere and you get another level of hot.

"Big" because it involves the entire visible universe.

And "bang" because the expansion that followed makes it look like there was an explosion in our past, right after our universe was born, although you will see later that it wasn't an explosion at all.

An "Extraordinarily, Tremendously, Mind-bogglingly, Beyond Blisteringly Scorching, Hot, Huge, Ubiquitous, Universal Deflagration" may convey the idea of what happened back then better, but "Hot Big Bang" is quite efficient too, and more humble.

And humble it should be, for even though it may seem to your cosmic eye that everything about this Big Bang is centred on our planet, Earth, it is not the case.

As you will now see, the Big Bang did not happen at one particular point in spacetime, but everywhere.

8 | *Beyond Our Cosmic Horizon*

When you were on the beach, at the very beginning of your journey, you wondered whether what you could see in the sky, with your naked eye, was the entire universe.

You now know it is not.

Our eyes only allow us to spot a few hundred stars, all belonging to our galaxy, the Milky Way, and some faint traces, for those who know where to look, of a few other close-by galaxies.

Using telescopes, the whole observable universe, you now know, is unfathomably bigger than that. But it too has a limit: the surface of last scattering.

This surface lies in our past, some 13.8 billion years ago.

But it also lies in space: about 13.8 billion light years away.*

It limits what we can see today.

Any light coming from further away would have had to travel for more than 13.8 billion years to reach us. But more than 13.8 billion years ago, light was not allowed to travel freely. It was stuck. The whole universe was too dense back then. Light only became free to shoot through space and time 13.8 billion years ago, and the last-scattering surface is the image that remains of that moment. Seen from there, it marks the beginning of a transparent spacetime. Seen from Earth, it marks the edge of the visible universe.

In a sense, this surface is our cosmic horizon. We cannot see further. Not from Earth anyway.

Since the beginning of this book, you've been travelling within the universe as it is seen from Earth.

You've always limited yourself to the visible universe, the uni-

---

* In fact, it lies much further away than that, because the universe has kept on expanding since the light that reaches us now left. Physicists estimate that distance now to be about 46 billion light years.

verse that lies within our cosmic horizon, a horizon that is centred on us.

But what about the universe as it is seen from somewhere else, from somewhere that is not Earth? Would the cosmic horizon there still be centred on Earth?

Picture yourself drifting on a raft, in the middle of the ocean, far away from all lands. The horizon is clearly visible to you: a line separating the water and the sky. Looking around, you can see that it forms a circle, a circle whose centre is *you*.

Does that mean you are at the centre of the ocean?

Of course not.

It means you are at the centre of the part of the ocean you can see, your *visible* ocean. There is no way you could see beyond its edge, beyond this horizon of yours.

But it doesn't mean that there is no such beyond.

There is.

Of course there is.

A friend drifting on another raft some distance away would also have a horizon surrounding her. *Her* horizon, delimiting *her* visible ocean.

If she is close enough, she could be within your visible reach. Your visible oceans would then have some waves in common, but she'd be able, in some directions, to see further than you, further than your horizon, and so would you of hers, in the opposite direction.

But she could also be beyond your horizon in the first place.

In that case, you could well have parts of your visible oceans in common without knowing of each other's drifting existence.

A third possibility is that your friend is so far away to start with that her visible ocean and yours have nothing in common at all. As seen from the sky, it'd mean that the circles limiting what you can each see do not intersect. Whatever could be seen from where she is would be entirely hidden from you. She could be watching some volcanic islands and whales, but you would know nothing of it.

In space, it is the same.

The visible universe that we can see from Earth is a sphere 13.8 billion light years in radius.

But that doesn't mean there is no beyond.

Someone else, on another planet, would be surrounded by his or her own cosmic horizon, which also would be 13.8 billion light years in radius, because there is no reason for the universe to be younger or older there compared to here.

The three cosmological principles you've heard of before have been introduced to ensure this: a visible universe so remote that it has no visible part in common with our own should look similar to ours (not identical, obviously, but similar), and obey the same physical laws.

Even if her raft were too far away for you to see her, you wouldn't expect your friend's visible ocean to involve flying mountains.

The same goes for outer space. The laws of nature should be the same everywhere. And no particular location should differ in this respect from any other.

It follows that the visible universe seen by anyone living anywhere in our entire universe (beyond the visible one) should also be expanding and obey Einstein's equation, meaning that if we were to rewind time we would find a Big Bang, just like here. A Big Bang centred on them, this time, not on us.

With such a vision of our *whole* universe, there is no such thing as a centre of it all, and the Big Bang occurred everywhere.

With such a vision, you get a taste of what one calls a *multiverse*: a universe made of many separate ones, unable to communicate with one another, although they all belong to the same whole.

You will see four different instances of such multiverses before the end of this book. This is only the first, and I introduce it first because most scientists do believe it to be correct.

Now, accepting this, does it mean that the whole universe, the "everything" one gets by patching together all the visible universes as seen from everywhere, is infinite?

No, it doesn't. The whole ocean, for instance, the ocean one gets by patching together all the visible ones seen from however many rafts you like, covers the Earth, which is finite.

The whole universe is finite, then?

No. It may well still be infinite.

We don't know.

As I mentioned at the beginning of the previous chapter, Einstein's equations unfortunately do not give us an answer to that question.

All right.

Now, what has been proven here? Not much, you think? Nothing, even?

Perhaps even the Big Bang theory seems weak to you, just an abstract thought.

Well, it is true that one could argue that what your friends saw in the sky (the further away faraway galaxies are from us, the faster they move away from us) merely indicates that the universe is *currently* growing. Many possible pasts could have led to such an expansion. No need to introduce all this Big Bang nonsense.

One could argue this, yes. But not for long.

Science is not politics.

Nature does not care much about anyone's opinion, even if they are in the majority.

Hard experimental proofs are always necessary.

And as we will now see, there are indeed some hard pieces of evidence for a Big Bang lying in our past, hints that are so compelling that some people go as far as to consider them proofs.

Were our universe (let's stick to the visible one) to have been smaller in the past, how could you prove it? Physically time travelling there is not an option, but you could *look* into the past.

By now, you should be used to the fact that when you gather the light coming from stars shining billions of light years away, you are seeing what they looked like many billions of years ago. You are watching the past. You can therefore check if the universe was smaller back then, or look for hints to this effect in the way light reaches you.

It is not always easy, however, to make sense of what one is seeing in the outer reaches of the universe. By far the best way to proceed is to have a firm picture of what to expect and then check whether that picture corresponds to reality. That is what theoretical physicists do (at least it is what they are supposed to do – sometimes).

But for now let us see what conclusion you can reach before looking up through a telescope.

You are back on your tropical-island beach.

It is deep into the night but, instead of watching the stars, and after checking twice that no one is around, you start talking to yourself, thinking out loud, to build a picture of the history of the universe in your mind . . .

"*If* the universe is expanding, then it must have been smaller in the past."

"OK."

"But *if* it was smaller in the past, then gravity, or the curvature of spacetime, must have been much more pronounced back then, since all its matter and energy must have been contained within a smaller volume."

"That is what Einstein's equations say, anyway."

"All right."

"Back then, spacetime grew because, for some reason, there was an expansion. It started out tiny and very densely filled with matter and energy, and then, after 13.8 billion years of expansion, it became what it is now, with planets like Earth and stars like those you can see above your island."

"*If* that is the correct picture, back when the universe was small . . ."

"Whether it was dense with mass or with energy actually doesn't make any difference, since mass and energy have the same effect on the geometry of spacetime. That's what Einstein also said."

"So far, so good."

"Now, *if* all that energy was contained in a tiny volume, then there surely must have been a lot of friction and other things happening, and it all must have been very hot, in the early universe."

Sounds fair? Yes, this is not the first time you've reached that conclusion.

But then, there are other conclusions one can reach from this.

Like this one: the universe could have been *so* dense that *none* of the light around at the time could have travelled through it.

"No light could have travelled through it . . . Hmm . . . That sounds like a wall . . ."

It does, you are right.

Well done.

Such a place *must* have existed at some stage in our universe's past if the expansion model is correct all the way through, and, well, such a place does exist. You've seen its surface. It is the surface of last scattering, the surface that limits what can be seen of our universe.

What you've just done is quite extraordinary.

You've just lived a physicists' dream: from pure logic, using Einstein's equations and what you've seen of the universe since you left your beach, you figured out that a wall opaque to light should exist out there in our past. That its surface should still be visible and . . . it is. This surface has been detected experimentally, and somehow mapped, as you will now see.

I understand that, while reading this, you may not feel like you've just revolutionized our vision of the universe, but that is because you have been introduced to that wall before thinking about it. You didn't spend twenty or so years of your life trying to prove that it should exist, long before it was seen. Those who did, however, felt incredible when the wall was proven to exist.

How was it proven?

Well, as you now start strolling the beach again, you realize that there is a problem: the surface that you saw at the edge of today's visible universe did not quite match the one you just thought of, did it? The real one, the one we can see through our telescopes, is very cold, whereas the wall you just pictured in your mind is supposed to have been very hot.

How hot?

Using Einstein's equation, some people actually did calculate its supposed temperature. And they came up with a rather large number: around 5,400°F. The whole universe, when it became transparent, must have been that hot, they discovered.

The wall you saw in the sky wasn't.

And that's a problem.

But haven't you forgotten something?

Didn't you suppose, to infer the very existence of a hot past, that spacetime expands, that the universe's visible volume has been growing up with time, to match what your friends saw in the sky? And could that expansion have an impact on the universe's temperature?

Well, yes. Not only could it, but it should, and that actually changes everything.

Take the oven you have in your kitchen. Heat it up so that the air within it is lovely and hot. Then switch it off and imagine that the oven undergoes a rapid growth to become the size of a building. The temperature inside will instantly be much lower than it was when it was tiny.

Calculations made as early as 1948 by US scientists George Gamow, Ralph Alpher and Robert Herman showed that, due to the universe's expansion, only a faint trace of the above-mentioned 5,400°F radiation should remain and fill our entire visible universe as if emanating from the surface of your wall. What temperature did they expect to find? Something around −436°F or −454°F. Something between 37°F and 55°F above absolute zero.

And it so happened that in 1965, some seventeen years after Gamow and his colleagues' guess, two US physicists called Arno Penzias and Robert Wilson found a peculiar job for the Bell laboratories in the US. They had to set up an antenna to receive radio waves echoing off a balloon satellite. A nice easy job, were it not for a rather strange hindrance, a troublesome noise they heard throughout their signals. To get rid of it, and get paid, they did some brilliant checks and looked for many potential engineering faults. But nothing worked. Whatever they did, the noise remained, unchecked. Eventually, finding no other reason for it, they blamed pigeons, or whatever birds were flying by, for having relieved themselves on their state-of-the-art antenna. Despite their impressive academic achievements, they went on to spend a long time furiously cleaning their device, cursing the existence of birds. But the noise failed to disappear, and they ended up calling some theoretical-physicist friends. Soon after, they realized that they could have tried to get rid of that noise for ever without the slightest chance of succeeding. What they were hearing was not due to some birds' gifts. The "noise" was not even from Earth. It was a signal. A signal with a temperature, a temperature of $-454.76°F$. And it was coming from space. From everywhere.

Gamow, Alpher and Herman had predicted it. It was a consequence of Einstein's equations. It was the leftover temperature of the last opaque moment of our universe; a freeze-frame snapshot of a moment over 13.8 billion years ago, when a much smaller universe was so densely filled with matter and energy that no light could yet travel through it.*

Penzias and Wilson had experimentally confirmed a prediction of a theory that seemed so absurd to some scientists that even its name, the Big Bang theory, was coined by one of the most renowned professors of the time, British scientist Fred Hoyle, from Cambridge University, just to ridicule it.

Penzias and Wilson received the Nobel Prize in Physics in 1978. They had discovered the heat that remains from the furnace our universe was a long time ago, the heat that radiates from the surface of last scattering, the surface that marks the end of the visible universe.† This radiation, one of the (hot) Big Bang's smoking guns, is called the *cosmic microwave background*.

---

* In case you're wondering: in a billion years, this surface will still be the same, but further away, and thus dimmer. And in hundreds of billions of years, it won't even be observable any more. So some day in the very, very far future, our descendants won't even be able to prove that our universe started with a Big Bang . . .

† You may also have been wondering for a while now why the surface of last scattering is thus called. Well, when light (a photon, say) hits an electron, it is said to scatter. Before the wall, light scattered off matter all the time. Matter was so densely packed that scatterings happened continuously and photons could not travel at all.

Penzias and Wilson had proved that the Big Bang theories were on the right path.

•

Now, why is this radiation called "microwave"?

That, again, is related to the expansion of the universe.

Light emitted at the time of the last scattering, when the universe became transparent, was very visible indeed and contained different colours and energies and frequencies. But it is no longer visible to our eyes – it got stretched.

Do you remember that the colour, and energy, of light waves depends on the distance between two successive crests? Well, stretched by the expansion of spacetime for 13.8 billion years, starting with the colour indigo, it gradually turns blue, then green, then yellow, then orange, then red, then . . . it becomes invisible to our eyes, and turns infrared, and then exists only as a microwave.

We are at the microwave stage today. What once was visible as hot, visible light has now become, after 13.8 billion years of expansion, −454.76°F cold microwave light.

With this realization, Big Bang theories were suddenly nothing to joke about any more.

But what do these theories mean? Do they suggest that the universe was created at the surface of last scattering?

No, they don't.

You've seen, in the last chapter, that the surface we, on Earth, see

---

Hence the opacity of the universe. But the universe expanded and became less dense. So much so that, one day, light could travel freely. That's when light scattered for the last time, making the last-scattering surface appear in our past. That is your wall. It is the light from that moment, light that we still receive today, that Penzias and Wilson detected, after it had travelled for 13.8 billion years.

at the end of our visible universe means nothing for observers who are not on Earth: they have their own.

But now what about us?

If the universe wasn't created there, there must be something beyond.

What might we find there? Do we know? Is that the Big Bang?

Well, in a way, yes.

The Big Bang lies behind the surface.

But not right behind.

It happened 380,000 years before.

380,000 years before the universe became transparent.

Behind (or beyond or before) the surface of last scattering, what later became our visible universe can be described as a soup of matter and light and energy and curvature that gets denser and hotter. You will soon be ready to travel there and see it all for yourself. But for now let's just say that the further you travel beyond the wall, into the deep past of our universe, the more extreme everything gets. Travelling too far beyond, you end up surrounded by nothing that makes sense any more. Even space and time end up being so curled up that Einstein's equations break down and can't keep track of what is happening.

When this happens, theoretical physicists reach a place where nothing can be said about anything any more. This moment can be considered to be the birth of space and time as we know them. According to the definition we shall use throughout this book, it is beyond the Big Bang itself.

To reach this place and figure out what the Big Bang is will be your mission in Part Five.

In Part Seven, for your final journey, you will go even further, beyond the very origins of space and time.

Why not go there right now?

Well, because for now you should take a few seconds to breathe and congratulate yourself.

You've come a long way since you first landed on the Moon. You've learnt many facts about the universe that your great-grand-parents would never even have imagined possible.

You've learnt that our universe's fabric is a mixture of space and time called spacetime and that not only is it shaped by what it contains, it also evolves according to its geometry and contents.

You've learnt that it is huge by any standards, even bigger than we can see, and that we do not know its shape, or its extent.

Our visible reality is large indeed, but it has not always been so.

You've learnt that the universe has a history, that it most proba-bly had a beginning, about 13.8 billion years in the past, hidden behind a surface opaque to light.

And you've learnt that it has expanded ever since, growing bigger by the minute.

And you should be proud of having figured all this out.

Why not go straight to our universe's beginning, then?

A good reason could be that you should try to figure out what our universe contains first. Without that knowledge, you stand no chance of ever unravelling its deepest secrets. Neither about its possible origin, nor about its possible fate.

"All right, let's do it!" you shout to yourself, opening your eyes.

A gentle night breeze blows over the ocean. The Moon is full. Its round surface reflects rays from the Sun, bathing your island in silver light and shadows. A few turtles shyly crawl out of the water to spend the night on the sand, and maybe to lay their eggs, if the day is right.

And you feel incredible.

"I'll be back!" you shout at the stars.

But you are not alone any more.

As you become aware of whispers behind you, you turn around, only to see your friends debating your situation with your great-aunt.

Having heard you talk to yourself on the beach all night, they decided to move your departure-date forward and looked for the earliest flight home. Your plane takes off in a couple of hours. You should pack, and get some rest, they say.

And your shouts and complaints and philosophical objections and speeches about freedom make no difference.

You are being sent home.

Now, however sad you may be about leaving the sea and the birds and the sweet wind, let me just tell you this: your journey through modern scientific knowledge has only just begun.

Part Three

Fast

1 | *Getting Ready*

Our senses are adapted to our scale, to our size, to our survival here on Earth. Our eyes are tuned to judge whether a fruit is ripe enough to eat, our ears to listen for danger and our skin to the cold of ice and the heat of fire. Our senses allow us to see, to smell, to touch, to taste and to hear our environment, this world, this reality within which we live.

But this reality is not the whole picture.

We are rather small compared to our planet. And the Earth itself, in turn, isn't much compared to the cosmos, as you have seen in your travels through the universe. It would therefore be rather strange, would it not, if, in order for us to survive on our modest little planet, our bodies had instead evolved spectacularly high-spec senses, able to register every known and unknown stimulus in the entire cosmos.

In the everyday life of a human on Earth, for the whole of human history thus far, our bodies simply have not needed to apprehend the mysteries of the subatomic world, the speed of light and the full array of light, from microwaves to X-rays. Indeed we cannot even tell the difference between two extremely hot temperatures, or for that matter two extremely cold ones: they would melt or freeze our fingers before we could judge their subtleties. It is much more important, for our survival, to pull our hand out of the fire, or to protect it from the cold.

We can detect the mild acid of a lemon with our tongue, judging whether it is good to eat; but we cannot judge the caustic difference between sulphuric and hydrochloric acid – they would burn a hole through our tongue.

Similarly, our bodies do not feel the curves of spacetime beyond their straightforward gravitational effect: as far as our everyday lives are concerned, all we need to know is that we are safely on the surface of our world.

The world we perceive through our senses, then, is by its very nature limited. Our senses are our windows on the world, but they are only tiny portholes looking out on an immense sea of darkness. And for millions of years, our intuition about what we confidently call our "reality" has had nothing but these sense-perceptions to build itself upon.

But that is not the case any more.

We can now see beyond our senses.

And beyond, reality changes.

In the first two parts of this book, you've travelled far and wide. You've crossed intergalactic voids and even glimpsed just how big this universe of ours is. You've discovered that what Newton thought was universal about gravity actually isn't. Gravity, as Einstein told us, is the result of a bending of spacetime. It is not a force.

Newton taught us how to use words and equations to describe and forecast the behaviour of the world we detect through our senses. Einstein, with his theory of general relativity, carried you beyond, and it was not your animal senses that allowed you to follow him there. It was your brain.

Using it, you discovered a law that merges space, time, matter and energy into a theory of gravity.

This was your first "beyond."

•

You are now about to enter two different beyonds, like an adventurer strolling newfound continents where nothing is familiar and where nothing can be taken for granted – not even the laws of nature.

The first of these two beyonds is the realm of the very fast, and the second, the richest of them all, that of the very small.

They will definitely look rather alien to you at first (and second, and third . . .), and I guarantee that your "common sense" will be screaming at you that what you encounter simply feels *wrong*, but remember: all the matter your body is made of belongs to these exotic lands. The fact is that you are built out of realities for which the rules of nature are very different from the ones we are used to experiencing while lying on a sun-lounger on a tropical beach. Only by a very strange mechanism does the reality we perceive day after day appear to us as it does.

You are sitting in seat 13A, next to a window. There are seventy-three passengers on the aeroplane. They all look normal – except for your neighbour. He looks weird. You try not to look at him and almost wish you had not asked to be seated away from your great-auntie. You've only been on board for a couple of minutes but you were the last passengers, and the plane is now ready to take off. Your holiday friends are waving you goodbye from the ground, visibly relieved to see you leave. You sigh, though. However eerie it felt, travelling through the universe was good fun. You're not that keen to be flying back home right now.

The engines thrust the whole winged apparatus up into the sky, up the spacetime slope our planet creates merely by being there. You are pushed against your seat and thus feel heavier than usual. You are actually experiencing gravity exactly as if you were seated not in a plane but on the surface of another planet, a planet whose gravity is stronger than Earth's.

Yearning for another interstellar journey, you close your eyes and begin to fire up your imagination.

A beautiful alien landscape appears in your mind, with strange trees and lakes and a sky with twin suns. You remember that in just the past few years, mankind has detected thousands of planets orbiting faraway stars, a handful of them potentially similar to Earth.

The humming of your plane's engines slowly lulls you to sleep, and you start dreaming about being somewhere far away, flying in a futuristic plane through a pink, double-starred alien sky. A distant voice reaches your ears. It says that your plane has reached its cruising altitude and will now accelerate to the unprecedented velocity of 99.999,999,999 per cent of the speed of light.

Some time later, as your plane starts its descent, a stewardess's voice wakes you up. A quick glance at your watch tells you that you've slept for eight hours. You stretch and yawn, slide open the window lid and look out. There's only one Sun. Its light rays are bouncing off morning clouds, giving them a pink hue not unlike the alien sky you were daydreaming about before falling asleep. Underneath the aeroplane, however, the surface of the Earth does not look as expected at all. A seemingly endless ocean spreads all the way to the horizon.

You are supposed to be landing home in less than a minute, but all you can see is open water . . . Dark thoughts make your spine shiver. Has your plane been hijacked? The other passengers seem pretty relaxed, including your great-auntie a few rows ahead of you, and your strange neighbour is asleep. So no, not a hijacking.

But still. Something is wrong.

Has the entire Earth become flooded while you slept?

You've read somewhere that about 10,000 years ago the oceans all around the globe were much deeper than they are today, covering much of the continents. Looking out the window, you wonder. Could you have travelled back in time, to wake up above a flooded Earth inhabited by long-extinct species? The idea makes you smile, but you can't shake an uncomfortable feeling that something isn't quite right.

You were asleep for about eight hours, it seems. And travelling. Anything could have happened to you or to your plane while you were out cold.

Throughout your life, you have probably, like everyone else, become used to waking up pretty much where you fell asleep. Now imagine you had never slept before, and you were to nod off for the first time in your life. You'd surely be pretty confused when waking up. The first thing you'd check would be where and when you are, as some of us still do in a panic when waking up away from home. In fact, away from home or not, most of us do systematically check the time when we open our eyes in the morning. Only on rather rare occasions – after a particularly good party, for instance – do we also check the place.

It so happens, however, that waking up in the same place in which you dozed off has never happened either to you or to anyone else. Ever. Earth does not stop moving when you sleep. Every hour that passes, Earth travels a little more than 500,000 miles around the centre of our galaxy. And so do you. That's the equivalent of about twenty trips around the planet. Every hour. No one minds, though, as long as their bed stays still beneath their body.

Were Earth, or just you, to also be travelling in time, however, it would be a different matter. But that is not possible. Time travel does not exist. Or does it?

As you stare out of your plane window at a vast city in the middle of an ocean, you realize that you are not about to land on the Earth you left.

Understandably, you start to panic and try to jump to your feet, but the seatbelt holds you down and the sound of the roaring engines covers your cries. You wave frantically at a flight attendant who frowns angrily and motions for you to be quiet – before reaching for his microphone to remind everyone that any disruption during descent and landing is still, in the year 2416, punishable by law.

You open your eyes wide.

What year did he say?

A second later, your plane lands on water, and starts to cruise through an alley of glass skyscrapers whose style of architecture is alien to you.

As you stare blankly out of the little window, you hear the stewardess speak again. With the smooth, professional tone used by flight crews the world over, she welcomes you home on June 4 of the year 2416, four centuries after your departure, three days ahead of the scheduled arrival. The time now is twenty-five past ten a.m. and the morning mist is expected to clear soon, to give rise to some bright, sunny spells. All passengers should expect temperatures about ten degrees above early twenty-first-century averages. Thank you for choosing McFly Airlines, a member of the Future Skies Alliance.

2416.

You glance at your smartphone. No signal. Typical. Fortunately, however, your watch still ticks. And it seems to be under the impression that you've only travelled for eight hours. Not 400 years.

Something is very, very wrong.

Is this a trick? Did your friends plan all this?

You check your flight ticket.

It's a flight home all right.

Have you been drugged?

Worse still: could this be real?

Is a debt collector going to be waiting for you at the airport to ask for 400 years of unpaid rent? And the person you recently arranged to go on a date with? And how about the milk you left in the fridge? Important practical questions race through your mind and make your head spin.

400 years in the future.

And whose future, at that? Certainly not your own, since your body does not seem to have aged at all in the eight hours since take-off. The future of your friends and family, then? The city you've just landed in definitely looks like no city or town from the century you've grown up in.

Time really does seem to have been fast-forwarded outside the plane, while you were sleeping.

But wait . . .

How could time *outside* the plane have raced ahead on fast-forward while time *inside* did not?

This sounds absurd.

But it does seem to be the case.

It actually is the case.

And your aeroplane's exceptional speed is to blame.

3 | *A Time of Our Own*

Speed changes everything. Even space and time.

A clock that is moving through space at a very fast speed does not tick at the same rate as a slow-moving watch gently attached to your wrist as you stroll on a tropical beach. The idea of a universal time – a godlike clock that could somehow sit outside our universe and measure, in one go, the movement of everything in it, how its evolution unfolds, how old it is and all that – does not exist.

What has just happened to you now, in your plane, illustrates this.

The time we humans experience seems to be the same for us all – "universal" – but we experience it as such only because when compared to light, none of us (even fighter pilots) is moving much more quickly or much more slowly than anyone else, which is very fortunate for watchmakers.

But even though our senses may not be able to feel it, the fact remains that if all of the people, animals and objects that happen to be on our planet's surface were assigned their own clock, it would keep time differently to all the other clocks. We all have a time of our own, attached uniquely to ourselves. Einstein figured that out ten years before publishing his theory of gravity, the general theory of relativity to which you were introduced in the previous part.

Back then, unable to secure any kind of position in any university whatsoever because nobody wanted him, twenty-something-year-old Einstein was making a living as a patent officer (assistant, really) in Bern, Switzerland. But this didn't prevent him from thinking.

In between evaluating patent applications, he tried to imagine how the world appears to moving objects, depending on their speed. He was looking for a theory of moving bodies. He wasn't yet obsessed by gravity. Nor by the universe as a whole. Just by how objects manage to move within it.

At the age of just twenty-six, in 1905, he published his work, and the whole scientific community soon realized that someone no one had ever heard of had made an extraordinary claim from a desk lost somewhere inside the Swiss Intellectual Property offices: that clocks did not always tick at the same rate. Rather, that the ticking of clocks depended on how they were moving relative to one another.

Better still, the theory this unknown young man was putting forward could predict the actual time-difference one should expect two travellers to experience, according to their relative speed.

This theory is called the theory of *special relativity*.

Let us imagine twins.

Two of them, since they usually come in pairs.

A couple of years after Einstein's publication, French physicist Paul Langevin calculated, using special relativity, that if one of the twins was sent in a rocket for a six-month round trip away from Earth at 99.995 per cent of the speed of light, the one who stayed on Earth would have to wait fifty years to see his sibling come back. So according to Einstein, six months lived by the one who left in the rocket ship should equate to fifty years for the one who stayed on Earth, and for the whole of humanity too: our planet would orbit the Sun fifty times during the travelling twin's voyage. Although they are twins, they would end up not having the same age at all, one being forty-nine years and six months older than the other. A rather astonishing claim.

When you heat up a metal bar, it expands and becomes longer. It is said to *dilate*. If you aim the heat carefully, it is possible to have only the bar dilate, rather than, say, the anvil it sits on – that is, its surroundings are unaffected.

According to Einstein's special relativity, a similar phenomenon happens with time. With a rocket shooting at 99.995 per cent the speed of light or a plane flying at 99.999,999,999 per cent of it, it is the rocket and the plane and everything they contain that move fast. Not their surroundings. So it is their times, and their times only, that are affected by their extreme speed relative to the world around them.

What Langevin's twins experienced, indeed what you have experienced flying in that extremely fast-moving plane of yours, is what scientists call *time-dilation*. The faster one travels, the more pronounced the time-dilation.

A very peculiar phenomenon.

But Einstein's special relativity also suggested something even harder to accept: it said that as your time undergoes a dilation, the very length of things should contract . . .

Now, since you were fast asleep in your plane when this happened, please allow me to offer you another ride in the world of the very fast.

You are about to see what our reality becomes when looked at while travelling at mind-bending velocities.

So let's forget about your plane for now, and even about gravity.

Picture yourself on Earth, in a spacesuit, and imagine you have a pair of rockets attached to your back, rockets that are so good that they never run out of fuel. You kiss your present life goodbye and get ready to shoot off. Spaceward.

Now off you go, hoping that no random space rocks are in your path.

You are not just a mind travelling through the history of our universe, but a mind *and* a body, as you were last time, embarking on a ride through empty space, just for the fun of it.

You're already in space.

You check your watch.

It ticks the way it's always ticked: one second passes every second, it seems, whatever that means.

The Earth is moving away behind you, but imagine that there is a huge clock hanging over it, a clock that you can always consult, wherever you are, showing the time ticking, say, in your great-auntie's house, telling you what time and day and year it is there.

Your thrusters are powerful.

You are at 87 per cent of the speed of light.

One second still passes every second for the watch on your wrist and the cells in your body, but everything around you starts to become strangely distorted.

You turn around to look at the clock above the Earth.

As one second passes on your watch, two pass on your home planet.

Uncanny.

Your personal ageing has halved compared to everybody still on Earth. But as far as your perception is concerned, a second always takes a second to pass. It is the Earth's clock that seems to be moving faster.

You keep going.

You are now moving at 98 per cent of the speed of light.

Five hours on Earth now take one of yours to pass.

You look ahead, at faraway galaxies.

Strangely, all these shining blobs of light that seemed extremely remote a blink of an eye ago now don't seem to be that faraway. It is as if galaxies out there had jumped closer. Five times closer, to be precise.

But that, surely, is not possible.

You look at your watch and tachymeter (a device that measures speed, just like in a car). You are now flying at 99.995 per cent of the speed of light, the speed Langevin gave to the rocket ship of one of his twins. It still isn't as fast as the plane you boarded, but even at 99.995 per cent of the speed of light, clocks on Earth now tick 100 times faster than yours. A whole day and night on your home planet takes just one minute and twenty-six seconds for you. One of your years is a century back there in your great-auntie's house. And those faraway galaxies ahead, those galaxies that are supposed to be millions of light years away, how come they suddenly seem so close? Surely they cannot be *that* close after only a couple of hours of travelling!

Still, they are.

A hundred times closer.

Their distance away from you has just shrunk by the same amount by which your time slowed down when compared to Earth's.

This is not at all like the expansion of the universe, though. That expansion occurs in the same way whichever direction you look.

Here, it is different. It only happens in the direction you are travelling.

And it depends on *you and you only*.

So forget about the universe. Just think about yourself and focus on what you see.

To your right and left, nothing seems to have changed. Same for above and below, where distant galaxies are still pretty much where they were before you started speeding up. But the galaxies ahead of you definitely are not. Facing them again, there is little doubting that something fishy is happening: it is not just time that seems to be subjected to dilation, lengths and distances have also apparently become . . . shortened? Contracted?

It certainly seems so. The whole universe looks to you as if you were peering at it through a distorted magnifying glass that shrinks the distances ahead, but not sideways.

You check your watch.

One second still ticks every second. And you are still accelerating and everything seems to get even more distorted. Understandably, you become confused and scared, so you make a huge U-turn to get back to Earth, which you expect to be extremely far away . . . but it is right there! Turning your head around, the galaxies you were heading for a moment ago are now back where they were: extremely far away! Whatever the direction you are travelling to at this amazing speed, everything ahead, however far, seems to be but a stone's throw away, whereas the other directions don't change . . .

A few minutes later, still confused, you shoot past the International Space Station, which moves around Earth at a completely insane rate. You check the ticking of your watch: still one second every second . . . You pass by an astronaut whose movements are accelerated 100,000 times. The hands of her wristwatch spin like mad. You *see* the difference between her time and yours! *You see her life unfold.* Ten hours pass on her watch as just one tiny fraction of a second passes on yours . . . And she moves accordingly . . . And so does the space station, and Earth, and everything around . . . And your rockets still fire their power, thrusting you past Earth. Faster and faster. Towards infinity and . . .

Half a second of your time passes and the astronaut is now back on Earth and a couple of blinks later she's dead and her children have grown and had children, and the Earth spins thousands of days and nights and years and you're too far away now to see anything of it any more.

A few seconds pass for you.

You keep accelerating.

No point in heading back to Earth now. You'd land in a future so remote that you'd probably feel like an antique, and surely would be treated as such.

The whole universe ahead of you keeps looking closer and closer, and flatter and flatter.

Sideways, still the same. Only ahead, in the direction you are travelling, do distortions occur.

You are still accelerating.

You are getting closer and closer to the speed of light, but something is not right again: although your rockets have thus far been happily propelling you faster and faster throughout space and time, your speed has not been increasing much lately.

Instead, it seems, your rocket's energy is turning into . . . mass.

Yes, you are sure of it. You are getting heavier by the minute.

Years of diets ruined by rockets.

Who would have thought?

"WAIT!" you shout, annoyed by this above anything else, and everything freezes.

You are up there, floating somewhere far away in space, probably millions of years in our future now, but frozen still. Rather handily, so is the whole universe. Nothing is moving.

You can relax for a moment.

Good.

Let's ponder together the three counterintuitive aspects of the high-speed trip you've just made.

Firstly, time flowed differently for you compared to anyone on Earth, including the astronaut (whose time is so close to that of the enormous clock hovering over your great-auntie's house that we can consider them identical here). The actual mechanical watches you were both wearing did not tick at the same rate at all and the faster you flew, the more pronounced the difference. That is the first change. It is strange, I agree, but it is so.

The second thing you experienced was that distances shrank ahead of you: what seemed very far when you were not moving fast became very close when you did. Which is also strange, I agree. But it is true nonetheless. It is called *length contraction*.

And the third is that eventually you became more and more massive. Which is annoying to say the least, though perhaps not as unexpected as the other two, considering that you now know that $E = mc^2$ – so let's look at this particular by-product of fast travel right away.

Nothing with any mass can reach the speed of light, let alone break that speed. That's a law. So the faster anything with mass is travelling, the more difficult it gets to accelerate it. To see what this means in practice, picture yourself flying so fast that adding only one mile an hour to your tachymeter would mean reaching the speed of light.

You then take a tennis ball out of your pocket, and throw it ahead of you. Let's say for the sake of argument that you throw it at 20 miles per hour.

On Earth, that'd be easy. But right now, it isn't. In fact, it is impossible. *Nothing* can move faster than light. So as you fly along at just 1 mile per hour short of light speed, your ball simply *couldn't* go 20 miles per hour faster.

Nothing prevents you from throwing the ball, that's true – but if the ball can't travel faster than the speed of light, then clearly something else will have to give as you hurl it into the void ahead of you. And the answer is given by our old friend $E = mc^2$: the extra energy you give the ball by throwing it forwards is turned into mass, since it cannot turn into speed.*

---

* To be more precise, at such extraordinary velocities, Einstein's equation needs some corrections (and Einstein is the one who found them), but the idea is basically the same.

You already knew that mass could be turned into energy (within stars, for instance), and you here have an instance of the opposite phenomenon: energy is turned into mass. And that wraps it: you've just learnt, thanks to Einstein's theory of special relativity, why you were becoming more and more massive before calling out and freezing everything.

Now let's turn to the other two problems of your high-speed journey: the dilation of time and the contraction of lengths.

Most people (me included) are both baffled and fascinated when confronted with the fact that there is no universal time. Our common sense, honed by millions of years of evolution on the surface of our tiny planet, intuitively rebels against the idea. But even though we can see its effects on us and around us, time is a rather abstract concept, an intangible flow of something that is entirely invisible. So despite its strangeness, we can surely cope with the idea of it not being as smooth as was once thought.

Space, on the other hand, is something we believe we are familiar with. But that is a mistake. We are not.

A metre is always a metre, you might think?

Well, that is not correct. It depends on who is looking at it.

Space and time are linked to one another: if time changes, distances also have to give.

Why does it *have to* be that way, you wonder?

Why do distances and lengths *have to* contract if time dilates?

The answer lies in the existence of nature's absolute, unbreakable speed limit: the speed of light.

Were distances not to contract, then you would already have violated this limit.

In outer space, light travels at around 186,000 miles per second.

An Earth-based observer watching you fly at 87 per cent the

speed of light would see you travel 162,000 miles in one of his seconds.

Flying that fast, however, you must remember that the seconds you experience are now different from his. At 87 per cent the speed of light, one second of yours equates to *two* seconds on Earth – and in those two seconds, the Earth-based observer sees you travel 323,000 miles. That's twice the distance he sees you cover in one second.

Nothing peculiar here, right?

Wrong. Because although you'd have travelled 323,000 miles in two of *his* seconds, only *one* of yours has passed.

That's 323,000 miles per second, as far as you are concerned.

The speed of light being 186,000 miles per second, you'd have smashed the universal record . . .

But this is forbidden. Not by the police, but by nature. Remember: *nothing* can travel faster than light. At the beginning of the twentieth century, many experiments had already established both that idea and the fact that in outer space, light always travels at that speed (no more, no less). Newton would never have been able to explain this with his vision of the world. But Einstein did. With his.

In his theory of moving bodies, the special theory of relativity, times and distances have to dilate and contract in such a way that, whoever is watching, no object can ever cross the light-speed limit from anyone's point of view.

An Earth-based observer's time flows twice as fast as yours? Then the distances you travel are, from your point of view, half the ones the observer sees you cross.

Flying at 87 per cent the speed of light you do *not* travel 323,000 miles every second, but 162,000. What seemed to be a mile for an Earth-based observer is actually half a mile for you.

Your speed is always the same, whoever measures it, be it you or anyone else.

Velocities are not observer-dependent. Only time and lengths are.

If, flying faster, distant galaxies seemed much closer to you, it is because they *became* much closer. For real. And this does not just apply to distances: objects themselves shrink with speed too. As seen by anyone not moving along with it, any rocket and all its passengers would contract. Even *you* did. At 87 per cent the speed of light, flying like Superman, fist stretched out ahead, you shrank to half your length as measured by someone on Earth. And someone flying with you would not have noticed it, for their tape measure would also have shrunk . . .

And this is all a consequence of accepting a fixed, finite and unbeatable speed for light.

All this is what Einstein wrapped up in his 1905 special theory of relativity: a theory that gives the rules of nature for anyone who cares to travel at (exceptionally) high velocities.

Strange? Yes.

Counterintuitive? Definitely.

But that's how nature works.

Now, what about gravity? We purposefully forgot about it for a while. But if we are to have a realistic vision of our universe, we now need to bring it back. In a moment, you'll therefore continue your fast-moving journey through a universe whose fabric, space-time, interacts with its energetic contents, and bends around them, creating gravity.

Back to you.

You are in outer space. Everything is still frozen.

Earth is somewhere far behind you. The astronaut you saw has been dead and buried for a very long time. You were shooting straight for faraway galaxies that now seem much closer.

Just remember that time and space are now inseparable parts of spacetime, the fabric of our universe, and that gravity is the effect of the bending of that fabric by the energy it contains, whatever its form, and that mass is energy.

You were getting more and more massive when you froze your journey.

Let's unfreeze the picture.

Ready?

You are flying again.

Your body is moving extraordinarily fast, and your thrusters are still powerfully hurling you forward. You are getting more and more massive and so, since gravity is back in the picture, your increasing mass is curving spacetime more and more around you.

The mass of a small mountain is now contained within your body.*

Rocks you fly by are beginning to slide down the slope you create and soon they even begin to start falling upon you.

They hurt when they hit you, but since you are getting more and more massive without growing in volume, you are denser than you were, so they shatter into tiny pieces.

Collecting more and more energy, you become as massive as the Earth.

You've captured big rocks and even small planets in your wake. They are orbiting you now.

You are so heavy, the curves you create in spacetime around your own body become so pronounced that the universe you see

---

* They may not reach the mass of a mountain, but that is exactly what happens to the particles accelerated in particle accelerators throughout the world: instead of reaching the speed of light, they gain mass.

becomes distorted in every direction. Not just ahead. And this is not due to speed any more, but to gravity, to the bending of space-time, to the energy that you have gathered within yourself. Because of this energy, space and time, intertwined as they are within the fabric of our universe, are so bent that wherever you look, the universe seems distorted, and accelerated, as if your time was now slower than every other clock in the universe.

You are about the mass of five Earths, all concentrated within your body. You are having a hard time lifting your hands, obviously, or anything else for that matter, and in fact now you can't move at all . . .

To be honest with you, if I were you, I'd stop right here.

Why?

Because sooner or later, accumulating more and more energy within your own body, you will end up becoming a black hole.

And that is not a good idea at all.

Unfortunately, you are already too massive to move, so you can't even check whether some hidden switch can shut your rockets off.

Your hands are now as good as glued to your hips and you are indeed starting to collapse in on yourself and . . .

"STOP!!!" you shout, in a panic, and you find yourself back in your plane, by your window.

Your weird neighbour is looking at you.

From his expression, it seems you woke him up.

He is definitely weird, but right now, you probably look even weirder.

You mumble an inaudible "sorry" and turn towards your plane window to stare outside.

It is dawn.

No sign of an immediate landing into a futuristic hometown.

No sign of faraway galaxies being closer than they should be.
No small planet orbiting you.
You are simply flying.
You stare at your watch.
It seems you've been in the air for eight hours.

"May I enquire why you cried out?" asks your weird neighbour.

"Where are we? What year is this?" you ask in turn, eyes wide open.

"I beg your—"

"What year?" you insist, rather nervously.

"2017!" replies the man, somewhat amused.

As the stewardess announces that the plane is about to begin its descent, you realize that you just dreamt it all, that you did not fly to the future, that you are still here, on your way to your lovely old normal hometown, with its concrete roads and brick-walled buildings.

The outside temperature is 53°F, continues the stewardess, and the morning mist will begin to clear by midday . . .

2017.

What a relief.

But what a strange dream.

4 | *How to Never Get Old*

What you just experienced was not a pure flight of fancy, though.

You really did get a glimpse of what the universe would look like if you could move very, very fast. Scientists have called velocities beyond which the strange effects you have experienced cannot be ignored any more *relativistic speeds*, and everything you've just dreamt obeyed the laws of nature, as they are understood today, from a relativistic perspective.

No human has ever reached these speeds, of course, but the particles that surround us have. In fact they do so all the time. But back in 1905, when Einstein came up with these amazing ideas, it was hard to check how they behaved.

It actually took sixty-six years after the theory of special relativity was published for two American scientists, Joseph Hafele and Richard Keating, to devise an experiment able to detect the strange time-dilation effects Einstein had forecast.

We are in 1971.

Hafele and Keating have acquired three atomic clocks, the best clocks ever made. Once synchronized with each other they remain synchronized to an extraordinary level of precision: they don't change by more than a billionth of a second over millions of years. Very, very reliable clocks indeed.

So, Hafele and Keating had three of them. Synchronized.

And they took them to an airport.

They kept one on the ground, in the airport lobby, and literally booked a seat for the other two on two commercial planes.

Just imagining the reaction of the other passengers makes me smile . . .

Anyway, both planes then took off. One flew eastward, the other westward, around Earth, before eventually landing back in their departure airport to join their Earthbound synchronized alter ego. Since Earth spins on itself in an eastward direction, flying eastward or westward does make a little difference to the overall relative speeds of the planes and airport.

Now, were nature to behave the way our intuition believes it does, the three atomic clocks should remain synchronized whatever the planes do. A second is a second, to the universal clock that God keeps on his bedside table, so one second should tick every second. All the clocks you've ever seen or used, mechanical or not, certainly do agree on that. And that's it. Except . . . no, that's not it. Nature doesn't care much about what our intuition believes and it so happens that this intuition is wrong. Our usual clocks are just not precise enough to tell us so. Our intuition may be wrong, Einstein's was not.

Once their two planes had landed back at the airport, Hafele and Keating found that their three atomic clocks were not synchronized any more.

The eastward-flying plane's clock was 59 billionths of a second late compared to the one that stayed in the airport. The westward one was 273 billionths of a second ahead.

It would have taken more than 300 million years for such a mismatch to happen naturally, had the three clocks remained next to each other.

•

According to Hafele and Keating, there were two reasons for this mismatch.

The first is related to the speeds involved, to special relativity: as Einstein had guessed, the relative velocities of the three clocks should indeed lead to some tiny – but measurable – time-dilation effects.

The other reason, however, has nothing to do with velocities, but with gravity, with Einstein's general theory of relativity: just as a heavy ball rolling on a rubber sheet bends the rubber more next to it than further away, Earth's effect on spacetime, Einstein said, should be more pronounced near its surface than far away in the sky, where planes fly, and will consequently affect the way time flows at different altitudes.

These two effects, independent of one another, were calculated *before* Hafele and Keating did their experiment.

And added up.

All in all, Einstein's theories predicted that compared to the Earthbound clock, the eastward-flying clock should end up being up to 60 billionths of a second late whereas the westward flying one should be about 275 billionths of a second ahead.

The experiment proved him right.

You may not find that so impressive because the time differences above sound tiny. And they are. But remember that a plane doesn't fly that fast, and Earth isn't that big a cosmic object. Fly faster and/or get close to a gravitationally much more powerful object in space, and the time difference can become huge, as you've experienced in the near-light-speed plane in your dream.

Needless to say, the precision of Hafele and Keating's experiment has been improved upon ever since 1971, confirming its result with increasing levels of accuracy. Spacetime indeed means what it means: it is a mix of space and time.

Within our universe, clock-ticking rates depend on who is looking at them: it depends upon where you are and what is next to you (that's the gravity bit) and on your speed. At the beginning of the twentieth century, this was very abstract. Today, it is an experimental fact. We all have to accept it.

In this universe of ours, time and distances are not universal concepts. They depend on the observer, on who is experiencing them and on who is looking at them. They are both relative. Otherwise, the speed of light would neither be fixed, nor a limiting one.

Now, what has mankind made out of this knowledge? Did it change our everyday life? The bit of it that is purely related to velocity did, yes – a lot. Not only does our technology often make use of fast-moving particles to convey information in all manner of transmissions, special relativity also helped us understand how all the matter we are made of works. As you shall soon see, electrons within the atoms that make up your body – and pretty much anything else in the world of the very small – move very quickly indeed.

As far as gravity and spacetime go, however, as amazing as it may sound, only one mass-market device has so far been built that makes use of their relationship: GPS. Any time you check your position with a GPS device, whether on your smart phone or in your car, you use the fact that space and time are curved around Earth. The closer you are to the surface, the steeper the curve – not only in space, but in time too.

There are clocks up there within the satellites that communicate with your GPS device to localize it. If no correction was made to take into account the time-ticking difference between the ground and the satellite, your position would quickly come out wrong. It would drift by about 6 miles per day. GPS would be useless. It is thanks to Einstein's special and general theories of relativity that GPS works.

All right. That's how it is. There is no such thing as a clock that ticks the same way throughout the universe.

Now, at 99.999,999,999 per cent of the speed of light, the plane in your dream travelled incredibly quickly through space compared to

Earth and all its inhabitants, and you landed in 2415, and you can consider yourself lucky.

Had you flown even faster, you would have arrived even further away in the future.

How much further? Again, it depends on your speed.

But there is one limit, because nothing can travel faster than light.

To travel *as* fast as light could maybe be possible one day, but you'd need to make a huge sacrifice: you'd have to get rid of your mass. All of it. Light cannot carry any mass whatsoever, and that is why it travels so fast. Light travels light.

What's the matter with matter? you may rightly wonder.

You experienced it yourself: everything that is massive becomes more massive when accelerated too much. To reach the speed of light, one must therefore have no mass to begin with.

Still, what would happen *if* you were able to transform yourself into a massless being of some sort? How, then, would your time flow? As shocking as it may sound, the answer is that it would not flow. The ticking of any clock you had with you (also turned massless) would just stop.

At the speed of light, time freezes.

Completely.

And that is the reason why the light that has travelled throughout the universe to reach us today is exactly the same as it was when it was emitted. Unlike a postcard that, after 13.8 billion years, would be tattered and ripped by the journey and resemble nothing close to what it once was, the images that are carried by light throughout the cosmos are not affected by the passage of time. When we gather light originating from the furthest reaches of our visible universe, we get pictures of the universe as it was back then.*

---

* Corrections due to the redshift induced by our universe's expansion need to be

Now, being made of mass, you have no choice but to be subjected to the passage of time. There is nothing you can do about that. To become eternal, you'd need to turn yourself into light, which can't be done. Still, if you could, then your time wouldn't flow any more at all. You'd be eternal indeed, but unaware of it. *

 That being said, despite the impossibility of being eternal, you may, even though you are massive (no offence meant), be able to reach a future unreachable to your neighbours. For this, you'd just have to travel fast, like your plane did. Or to settle on a gravitationally much more powerful planet than Earth.

To wrap up the subject, I know that some of you, for one reason or another, do not like the idea of ageing and would very much like to remain young for as long as possible, or at least much longer than your neighbour. Well, to you, special readers, I now address the following warning: there is no point trying to run quickly or to become a Formula One driver or even a test pilot for the US Air Force. There is even no point trying to board a plane to fly at 99.999,999,999 per cent of the speed of light.

 Why?

 Because *your* clock won't ever change from *your* point of view.

 To your eyes, and in terms of the cells that make up your body, a second will for ever and always be a second, a day a day, a year a year, and so on. Your personal time and ageing won't slow down, you won't live longer, your cells will still grow and decay at the same rate and that will be the same for everybody travelling with you. Travelling fast or living on a much denser faraway planet won't make you live longer, because to you twenty-four hours will still feel like (and be) twenty-four hours. However, *other people* may see you live longer than they do.

---

taken into account, though. Pictures we receive from the cosmos have been stretched by the universe's expansion, but they haven't aged.

Fast-forwarding your present to quickly reach another person's future is theoretically possible (and may even also be in practice some day),* but living longer by travelling fast isn't.

With Einstein's special and general relativities, you've discovered that a truly weird world surrounds the one we have access to through our senses, the one we live our daily lives in. But what you've seen so far is nowhere near as weird as what you are about to experience once safely back in your home.

After the very large and very fast, it is now time for you to enter the world of the very small.

And I am afraid that if you did not believe in magic before, you might have to start now.

---

* Coming back, however, isn't. So, if given the chance, do please think twice before signing up for such a trip.

Part Four

A Dive into the Quantum World

1 | *A Lump of Gold and a Magnet*

Your great-aunt has left. You did ask her to stay for a few days, if only to have someone to discuss your strange relativistic dream with, but – rather unexpectedly – she declined the offer. All things considered, she found you well and sound, felt she had played her part in bringing you back home, and boarded the first flight to Sydney, leaving in your care the whole collection of crystal vases she had brought to cheer you up. She's back in Australia now, and you are back home. On your sofa. Looking at her awful vases while toying with a small magnet in the shape of a palm tree that you bought in a souvenir shop, to remind you of your tropical island.

You still have a week before getting back to work, seven days to find as many ways as you can of ridding yourself of all those vases, but you hesitate.

Have you finished with your adventures in the hidden nature of reality, or is there another level of understanding still to come?

Finding no straightforward answer, you stand up to make yourself a hot drink.

Pottering about in the kitchen making coffee, you suddenly spot a brick that, curiously, sticks out ever so slightly from the wall. Surprised, you pull on it – and it slides out. To your astonishment, behind it lies a cube of gold, probably hidden there by some (really rather careless) former tenants. It is half the size of your palm and thus worth a small fortune. How you could have not spotted the brick before is a mystery, but there are arguably no better homecomings than those when you find some gold in your kitchen, so you don't think about it too much. You pour yourself some coffee and look at your treasure with a cunning smile.

You have journeyed through the cosmos, the realm of the very big.

You've travelled fast, as fast as it is possible to go.

But you do not have a clue about the world of the very small: what matter is actually *made* of. Is gold made of small bricks?

Why are the materials that surround you so different one from another? Why is gold different from cheese? Why aren't we liquid at room temperature like water is?

With a grin, deciding to put science before money, you cut your gold into two equal parts, to see what is inside.

Unlike some cheeses (but not all), the inner parts of the gold bar have the same colour, the same absence of smell, the same everything as its outer surfaces. You nonetheless cut one of the halves in two again, and again, and again, frantically looking for some change as you get smaller and smaller chunks of it.

It seems to be gold all the way.

One might think that this cutting could go on for ever but no, it does not. After twenty-six or twenty-seven halvings, you are left with the tiniest possible piece of gold there is: cut it once more and you still get something, but it is not gold any more.

This elementary amount of gold, the tiniest thing that is still gold, is what scientists call an *atom* of gold.

Note that although it may seem that halving something twenty-six times isn't much, it is. You would find it rather hard to do it at home. To give you an idea, were you to do it the other way round by, for instance, tearing off a page of this book before folding it to double its thickness twenty-six times, you'd get a stack about 9 miles high. Another way of expressing this is to say that you'd need a mountain 50 per cent higher than Mount Everest to start with if, by halving it twenty-six times, you wanted to end up with something as thin as a page of this book.

Only the best modern technologies are able to look at a single atom of gold.*

What about lead, or silver, or carbon?

Any other pure material you'd have found instead of gold would have led you to the same conclusion: halve a piece of it that fits in your palm again and again and again, twenty-six times – give or take one or two – and you end up with an atom, something that cannot be broken any more, without becoming something different from the material you started with. Cheese, on the other hand, is not a pure material. But it *is* made of atoms too, atoms that are stuck to each other. All the matter we know of in our universe is made of atoms.

So what are the atoms themselves made of?

You cannot tell yet, but you have a hunch that they are full of smaller constituents, and that these tiny pieces are the same in all the atoms throughout the universe. You will travel within their world very soon, but I can already say that it is because the number of these smaller constituents differs from one atom to the next that pure materials have very different properties and thus, as everyone knows, value. Any broker would certainly wonder about your sanity were you to try to swap two pounds of mercury (worth about \$35) for one of gold (about \$40,000) or plutonium (worth around \$4 million, depending on the market) on the basis that they're all made out of similarly structured atoms.

So what are they, these atoms? What gives materials made out of them such different properties and shapes? And why can you cut through butter with a knife, but not through a diamond, if everything is made out of the same stuff?

With all these questions stacking up in your brain, you approach your fridge to fetch some milk for your coffee and mindlessly grab

---

* And you'll hear about one such technology in two chapters.

your palm-tree magnet to stick it to your refrigerator, but as it leaps out of your fingers to settle itself right against the metallic door, you freeze.

Until now, such an action, for a magnet, was familiar enough.

But it is not any more.

How *do* magnets do that?

How does the fridge know that the magnet is coming? Or is it the magnet that knows the fridge is there? Or both? Or is it just pure magic?

As far as you know, you've never ever seen anything being exchanged between a magnet and a fridge. No spooky hand reaching out from one to grab the other and pull it against its surface.

But maybe you simply did not look carefully enough.

You unstick the magnet from the fridge and stare at its surface, on the back of the crudely crafted palm tree. As far as you can tell, the dark surface is flat.

Focused, now holding it very tightly between your thumb and index finger, you press your cheek against the fridge door, staring carefully at the air, and bring the magnet closer to it again.

It is a few centimetres away.

You feel something.

A force.

An attractive force is pulling the magnet towards the fridge. Or the fridge towards the magnet. Or both. Hard to tell.

But in the air, nothing. That's for sure. You can't see the slightest hint of anything happening that could explain how they are aware of each other's presence.

The magnet is now about half a centimetre away from the fridge and the pulling force is getting much, much stronger.

You even have a hard time keeping the magnet where it is.

And still nothing visible.

You let go. The magnet jumps from your fingers to the door, where it sticks still, as happily settled as you are curiously puzzled.

For centuries, many men and women have wondered about this strange attraction. It is spooky, isn't it? The magnet jumped. Nothing happened *before* it touched the fridge and yet there was a force. That's what our ancestors thought, when they looked at magnets, so even though they did not have fridges, they began to talk about a spooky *action at a distance* to describe the invisible something that makes magnets work.

It is a bit like gravity, really.

No one can *see* gravity.

When Newton came up with his amazing formula to describe how objects, throughout the universe, are attracted towards each other, he had no clue *what* was responsible for the gravitational force he was describing. Einstein found out, though, about a century ago. Gravity is not a force, he told us, but a fall. A fall down spacetime curves.

So is it the same with magnets? Do magnets create steep curves in spacetime too?

No. That can't be it. Or everything (wood, we ourselves, beer, anything really) would fall towards magnets, not just nails, iron filings and other potential magnets. You've never felt your own fingers being pulled by a magnet. No, something else has to be found. And something else has been found. About eighty years ago. It involves what we call a *field*. A quantum field, to be precise. And now that you know about the existence of atoms and magnets, you are about to see what a marvel a quantum field is.

Imagine for a moment that you are a fish, and that for some reason you've decided to see what lies above the ocean that is your home. From the depths, gathering as much speed as you can, you shoot upwards, like a torpedo. You are aiming for what we humans call the surface but what you, a fish, probably call the ceiling.

You swim fast. Then faster. Water slides against your scales. The surrounding light gets brighter and brighter as you move closer to the end of your liquid world. Presently, you're out. There's no more water around you. You are flying through some blue emptiness (we humans call it the atmosphere). You flip your fins as hard as you can, but there's no way you can swim higher. Unlike a bird, indeed very much like a fish, your voyage upwards comes to an abrupt end. Slipping, sliding down the spacetime slope created by Earth's presence, you splash back into the ocean.

Some time later, back in the salty depths of your liquid home, you discuss your experience with some fish friends of yours who share a similar taste for the unknown. You promptly agree that up there, above the ceiling of your immense liquid world, it is impossible to swim. Above the ocean, you conclude, there is only blue emptiness.

We humans know better. We know there is air above the ocean and we now know that what we call air is far from being nothing. Deprived of it for more than a couple of minutes, we die.

Most of us, however, are not much wiser than the fish under the sea: don't we all kind of think that in outer space, above the atmosphere, beyond our precious air, there is nothing at all? Don't we believe that space is but black emptiness?

As you are about to see throughout the rest of this book, that is a mistake.

Outer space is far from empty.

When, as a fish, you briefly jumped above the ocean's surface, you entered another world, a world mostly made of gas and dust, not liquid.

Now, the world you are about to enter is much more widespread than that. It is called the *quantum* world; it is the world of fundamental matter and light.

Unlike the sea, which is made of water and stops where air begins, the quantum world is everywhere. In the sea, in the ground, in the matter we are made of, in light, and in outer space. Even in "empty" space. To enter its realm, however, has taken mankind millennia. The doors to the quantum world are buried deep within the very small. And because the air and gravity and many other things can mess up the picture, we are about to forget about them for a moment.

And the best way to do that is to send you back into outer space.

As you pull your magnet away from the fridge door to check its surface again, you see no visible change there whatsoever. Still black. Still smooth. And yet you felt the force. No doubt about that. How weird.

You press your cheek against your fridge once more, to do the experiment one more time, and you are so focused on it that everything but the magnet and the fridge vanishes around you. The floor, the air, your lump of gold, the walls, your whole kitchen and your flat. Gone. Your hometown. Gone. And so is Earth and the Moon and everything else.

You are floating in outer space, in a world of thoughts that obeys the laws of nature as they are known today. There is no air around you. There is no gravity either. There is nothing, really, apart from you, the magnet, the fridge, and whatever it is that makes magnets and fridges interact.

By now, you should be used to this kind of situation, so you do not worry too much about it and focus on the task at hand.

Your cheek is cold against the fridge's door. The magnet is still in your hand. You set it loose and, at the very moment you let it go, a new kind of adventure begins for you: you start shrinking! Throughout your journeys through spacetime, you looked upon the universe from a rather large perspective to get to grips with the very big, and then you had to see the world from the point of view of extreme speed, so you travelled extremely quickly. Now you are on your way to discover the quantum world, so you shrink.

A lot.

You are becoming a mini-you. A mini-you who is a couple of atoms tall.

How small is that?

Let's see.

As you are reading this, your book, or screen, is probably just a few hands' breadths away from your eyes. The smallest thing your sight can perceive from that distance is about a twentieth of a millimetre thick, a third of the width of a human hair.

Right now, your mini-you has shrunk to a size 100,000 times smaller than that. About the right scale to see if there actually is anything happening in between your magnet and your fridge.

Focused, although somewhat taken aback by the shrinking, you look around for spooky hands reaching out from one direction or the other. You turn your mini-head left and right and up and down.

You do not see anything at all.

You know the magnet is somewhere to your right, and the fridge somewhere behind your left ear, but from your new perspective, they are too far away for you to see them.

So you wait.

And nothing is happening.

Nothing at all.

After a rather long moment of pure solitude, you decide to try something else: rather than seeing, *feeling* might do the trick. Like when you were a kid and, to kill time, pretended you had super-powers.

You virtually breathe in and out a few times, for concentration, and then you switch off your sight. You are like a minuscule yogi in space. Tinier than dust. Eyes closed, you slowly spread your arms as you've seen people do in movies.

At first, you do not feel anything. And then, you do.

You have the impression of being a fish in the sea, as if everything around you was bathed in some kind of . . . of what? Not water, obviously . . . You open your minuscule eyes, eager to see what that sea is made of, but the feeling immediately snaps away and again there is nothing around you. The impression is very strange indeed. A bit scary, even, but you are no coward and you quickly reason that just like so many other things in our universe, what you just felt is real, but invisible to the eye.

So you close yours again, to enter the quantum world, yogi-style. The "sea" is there, all around you. There are even . . . currents? Yes. It seems so. Originating where the magnet is supposed to be and ending on the fridge. There are loops of force around and they flow right through you, and you realize that what you are sensing is what makes magnets and fridges interact, it is the so-called *electromagnetic force field*. Behind your closed mini-eyes, it appears like a mist of strength that spreads all around, everywhere, a mist that is most dense next to the magnet and next to the fridge. Ripples are propagating through it, at the speed of light, telling you that the magnet and the fridge are moving closer to each other, meaning that they will sooner or later hit each other, meaning that . . . You open your eyes and stare in open-mouthed horror at the enormous black magnet that is about to crush you.

You step back, shaking with fright.

You are so close to the magnet now that you can almost see the atoms wiggling in its surface. There even seem to be tiny currents flowing within it. What are they made of, are they electric? Magnetic? Both? You have no idea, but what is certain is that . . . WAIT! WHAT WAS THAT?

Something happened.

You saw it.

It wasn't an arm reaching out from the magnet towards the fridge, but some light. Virtual or real, it is hard to say, but there was light. It popped out of nowhere, right in front of your mini-eyes, from above the magnet's surface. Or was it from within? You turn your head towards where it went and you see the fridge's door, immense, moving towards you as well . . .

You hold your mini-breath.

You are an instant away from getting crushed.

More and more strange pearls of light appear from within the emptiness that seemed to separate the magnet and the fridge a moment ago, an emptiness that definitely does not look empty any more. Pearls of light flash around you, exchanged between the magnet and the fridge, like a host of tiny angels dragging the two objects towards one another.

Mesmerized by the show, certain that your mini-body is living its last moments, you wonder whether these particles of light are products of your imagination or real . . . They do seem virtual, for they last but an instant and appear out of nowhere, but they also have a very concrete effect on the magnet . . . Yes, these bright little fellows carry the force that brought the magnet close to the fridge in your house . . .

You close your mini-eyes.

You are about to get crushed.

But *bing!*

You are back in your kitchen, staring blankly at the fridge's door, where the magnet just glued itself to the fridge with a tiny metallic sound.

Wiping off a drop of cold sweat dangling on your forehead, you breathe, slightly embarrassed – even though you're alone – for having thought that it was anything but your imagination.

It felt very real, though.

You've just witnessed an *action at a distance* that is not due to magic, although I admit it is rather spooky. You've seen, beyond any doubt, that the mysterious force that makes two magnets interact, the electromagnetic force, was carried by virtual particles of light, particles so strange they exist for one purpose only: to carry the electromagnetic force. They appeared in between the magnet and your fridge, out of what seemed to be nothing, but is not. You have just discovered that in between any two objects in the entire universe, whether magnets or not, there exists something, something that is called the electromagnetic field. A sea of force out of which virtual particles of light can pop at any moment.

Right now, as you stare at your fridge, countless such virtual little pearls of light are exchanged between the magnet and the door, but you can't see them any more and never will. That is why they are called virtual. They pop out of a void that is not a void, and disappear without allowing anyone to see them.

Such virtual force carriers are everywhere around you, right now, and even inside you.

They all belong to the electromagnetic field, an invisible mist that fills not just the space in between fridges and magnets, but the entire universe.

What about magnets that repel each other? You've certainly seen that, haven't you?

As you shall soon experience when flying through an atom, the virtual pearls of light you've just been introduced to can either attract, repel, or do nothing to the matter we are made of, to the matter that surrounds us. It all depends on what the matter in question contains. In fact, it so happens that it depends on one thing only: a thing that scientists have called the *electromagnetic charge*. And just as you can measure your mass on a weighing scale, you can measure your charge using an instrument, too. Yours, overall, is zero though – the human body is electromagnetically neutral (otherwise magnets would stick to you, which would be rather bothersome). But that is not the case for the individual particles that make up your body.

Only two types of electromagnetic charges can be found in nature. For convenience, they have been called positive and negative, plus and minus.

The rule is that virtual pearls of light repel *like* (that is, similar) charges, whereas they attract opposite ones. Plus and plus, just like minus and minus, are pushed away from one another by the virtual light that appears between them. And the closer they are, the more the virtual pearls of light, the stronger the repelling force. Plus and minus, on the other hand, like to cuddle. Like your magnet and your fridge. And the closer they are, the stronger they attract each other. Neutral objects, on the other hand, don't care about these pearls of light, and one can be neutral either by having exactly as many positive and negative charges (your body is like that), or by carrying no electric charge at all (some particles you will meet later on don't carry any). These are the rules of the electromagnetic field.

Now, you may think that, since such an explanation of how magnets and fridges interact cannot be seen with our eyes, this may all be a very useful mental construction but not quite something that corresponds to how nature really works. The electromagnetic field, you may then argue, is but a picture that gives scientists a way to describe how charged objects react to a magnet's presence. A picture. A clever and imaginative one, for sure. But nothing more.

You could think all this, obviously, but you would be wrong.

The field you've just been introduced to, this invisible mist that permeates the entire universe and becomes somewhat more active near and in between charged objects, is much more than that.

For one, it is very real.

In fact, not only does it rule everything that has an electric or a magnetic charge, it is also the entity that gives birth, everywhere in the universe, to every and any electromagnetically charged particle, as well as to light. The electrons you will soon encounter are expressions of it. The light that your eyes detect is another. They are both but ripples in the field.

Many amongst the brightest scientists on Earth today consider the electromagnetic field to be more fundamental than magnets themselves. Or even than fridges. More fundamental than light, even. And more fundamental than you. However absurd that last bit may sound.

Before the end of this part, you will become aware of the existence of two other quantum fields that also fill the entire universe. And you will realize that, as far as modern science is concerned, you and I and all the matter we know and see and all the light that shines anywhere are all but expressions, ripples again, of these fields. We humans are indeed like fish in a sea. A sea made of fields. Just like everything else. And even though our ancestors once lived in the sea, it still took them eons to evolve and figure out the existence of quantum fields.

3 | *Entering the Atom*

You've been blankly staring at your tacky fridge magnet for long enough. You shake your head and open your fridge to pick up that milk you so badly wanted before the magnet drew your attention to a rather spooky phenomenon.

Walking back to the table where you left your mug, you are about to pour the milk when the sight of the gold lying nearby makes you stop.

What exactly are those atoms of gold that you found earlier, or the atoms wiggling on your magnet's surface? Are they like small, round balls? Are they cubes? What exactly are the charges that the virtual pearls of light from the electromagnetic field like to act upon? And what on Earth did I mean by saying that they all are expressions of some fields?

As you might have expected, these questions send you right back into your mini-you state, and you find yourself floating in the middle of the kitchen, far away from any familiar object, curious to see what this atom of gold you singled out earlier is made of.

But it is not an atom of gold that you encounter first. Rather, it is the smallest atom there is. The one that makes up 74 per cent of all the known matter in the universe: hydrogen. The very atom that stars like the Sun fuse in their core to create bigger ones and, as a by-product of this fusion, to shine.

To be fair, however, you do not see much.

There is *something* in front of you, of that you are certain, but you have a very hard time finding out *where* it is, not to mention *what* it is. Sharpening your mini-eyes to focus even more closely makes no difference at all, so you decide again to try to *feel* it, yogi-style.

Amazingly, it works.

Your eyes are closed, but you can picture something.

Some kind of wave rippling the ambient electromagnetic field . . . a wave wiggling around a sphere . . . a hollow sphere, or a hollow lobe, rather . . . and it is not really a wave, in fact . . . but it looks like a spherical, no, lobe-shaped one, with ripples that move fast . . . at a speed very close to the speed of light, so that the world it sees must be very distorted, not to mention the way its time ticks compared to yours, but it isn't really focused in a particular position . . . All right, let's be frank, you do not know what you are picturing, but this whole spherical, or lobe-shaped or whatever, fast-moving* thing does carry an electric charge. You can feel its effect on the background of the electromagnetic field, just like you did with the approaching magnet.

Is that what an atom is?

Still focused, you realize that there is something else . . . Something buried deep down, something very small compared to the volume spanned by the moving wave, but something that has to be strong, very strong, even, to keep that moving charge you feel from wandering away.

The hydrogen atom, you realize, has a core surrounded by a moving charge. All atoms in the universe have this structure: a core of varying size surrounded by one or more electrically charged waves.

Scientists have called that core the atomic *nucleus*, while the fuzzy, charged, wiggling wave is an *electron*.

And this is a puzzling revelation.

The electron looks nothing like the tiny dot you had pictured it to be.

To make sure you aren't getting this wrong, though, you quit your yogi mode and open your eyes. Rather unexpectedly, the wig-

---

* Fast-moving, here, can even mean relativistic, i.e. moving at a significant fraction of the speed of light.

gling wave immediately disappears, to become something else, something that looks much more like a particle.

Good.

Electrons exactly identical to this one are present in various numbers in all the atoms of the universe. They are the basis of all our electrical and magnetic devices, be it a computer, a washing machine, a cell phone, a light bulb . . . anything. All our energy and communication tools depend on them.

Slowly, very slowly, you therefore move one of your tiny hands forward, to grab this one and study it up close.

Strangely enough, the electron is very hard to catch. Every time you manage to spot it from the corner of your mini-eyes, it starts moving erratically, as if the very act of your trying to locate it made it change its course in an unpredictable way.

This is not your imagination playing a trick on you.

It is a real phenomenon. One of many that occur in the quantum world but not in our everyday world of crystal vases and cups of coffee.

It is part of a fundamental uncertainty of nature as it is seen from our point of view. You will have a deep look at what it means in Part Six, but you already feel that there is something uncanny going on. What you need, you think, is to actually capture this electron – and make it talk. That's right. Mini-you or not, you are pure mind down here; you can do whatever you please. And you'll be damned if a tiny electron will prove you wrong, so . . . hop! Faster than thought, as your mini-eye catches a glimpse of it, right there, to your right, you pounce on it. And there it is now, in your right hand, which you keep tightly closed. The electron is wiggling inside; it feels as if a butterfly flying at nearly the speed of light is flapping its wings against your palm. You start to squeeze your fingers. Electrons are charged particles; they interact with the ones

contained in your mini-hand through virtual pearls of light popping out of the electromagnetic field.

You squeeze, and squeeze, and squeeze, eager to have it quieten
down within its tiniest of prisons and . . . suddenly, you don't feel
it any more. It is gone.

You open your fist.

No electron in there.

You are absolutely certain that you did not leave any gaps
between your tiny fingers but still, it leaped out. And you felt nothing. It jumped through you, without touching you.

It is back surrounding the invisible core of the hydrogen atom
you had taken it away from.

How rude.

But how did it do that? How could the electron leave your grip
without touching you? Well, it tunnelled through your hand. It
jumped. A record-breaking jump. A quantum jump. Something
that is confined to the subatomic world and that doesn't exist in
everyday life at the macro level of kitchens and vases and aero-
planes. Or so one might think.

You haven't managed to analyse an electron yet, but you already
know one of its weird properties: it can jump like nobody's business. The phenomenon itself is called *quantum tunnelling* or *quantum
jumps*, and it so happens that not just the electrons but *all* the particles you will find in the quantum world can quantum jump – or
tunnel – like that.

Now that we have established that, let's pause for a second to reflect
together upon terminology.

When scientists discover something new, they need to give it a
name. For the very small, for the quantum world, they make up
word-associations where the word "quantum" is followed by

another, usually from the vernacular. Here we have "tunnelling," or "jump," or "world," all terms that are easily understood and which, on their own, do mean what they mean for us in the everyday world. The presence of the word "quantum," however, serves as a warning. "Quantum" means there is something fishy going on. In the example at hand, the fishy thing about quantum tunnelling is this: electrons really do tunnel through things . . . but there is no tunnel.

Quantum jumps hardly ever happen at the human scale, but imagine if they did. Imagine you are back in time, when you were a kid, in this very kitchen. Your father just asked you to clear the table, but it is late and you suddenly feel all of the 62 miles of air that weigh on your frail shoulders. You mumble something barely audible but not unlike a bear cub's growl. Nothing works though. The table is waiting for you.

You sit down on the floor, full of despair. And there you go. You suddenly find yourself in the dining room, on the other side of the kitchen wall near the table, and everything, all the cutlery, the plates and glasses, they all tunnel or jump or whatever right through the wall to the kitchen. This might sound like a fairytale or a scene from *Mary Poppins*, but to be fair, with quantum jumps like this, there's no telling where the cutlery and dishes and glasses would jump to, so there's hardly any chance at all they'd end up in the dishwasher, and your father would have to buy everything again, for you'd never find any of them.

Sounds strange, doesn't it?

Well, that is quantum tunnelling. Doors and walls and privacy would not exist were quantum laws to apply at our scale. Fortunately, and rather mysteriously, they don't.

Thanks to quantum tunnelling, however, almost everything in the realm of the very small can cross any barrier. How? It is understood that they can do so because they are allowed to borrow energy from the quantum field they belong to, the sea within which they swim, a sea that fills every single place in spacetime. As much as they want. The dream of all athletes.

But that doesn't tell you what an electron *looks* like, and I'd rather be quite honest with you: your mini-you may have to face a slight disappointment here. Picturing an electron is not going to be possible, because of the very quantum field it belongs to.

The electromagnetic field is everywhere, and every single electron that exists in the universe not only belongs to it, but also is exactly identical to any other electron, anywhere and anywhen. Interchange two of them, and the universe won't notice. Because of that, because of the quantum field they are an expression of, electrons cannot be described as one would describe a macroscopic object. They belong to the field. They are part of it, like a drop of water in the vast ocean, or a gust of wind in the night air, a drop or a gust you cannot localize. As long as one does not look, drops and gusts are just like the ocean itself, like the wind. Mingled into an entity much vaster than themselves, they have no identity of their own.

In the quantum world, as soon as you look, the electrons do become particles with given properties, like drops taken out of the ocean, but their properties are like nothing you've ever seen. They do not behave as expected – or at least as our senses might expect from our experience of everyday life.

If you know where an electron is, you *cannot* know how fast it is moving: its speed becomes unpredictable. That is why you had a hard time finding the electron around the hydrogen atom. Anytime you saw it, it started to move erratically. You were unable to follow it and it vanished from your sight.

In a similar way, if you know how much energy an electron has, you *cannot* know how long it will keep it.

Energy and time, position and velocity, are concepts that are not independent of each other down there within fields, in the quantum world. You will hear more about all this in Part Six, but for now, since your mini-you is touring the quantum world for the first time, you can consider it a warning (and maybe a teaser for some). Your mini-you should just take it all like you did as a young child discovering the world: without prejudice. Position and velocity cannot be known simultaneously? Fine. That's the way it is. Quantum laws allow for otherworldly jumps and tunnelling? Okay, so be it. Interpretation will come in due course, or not.

That being said, this whole quantum-tunnelling business does sound like utter nonsense to me as well, and I have been told that, after a course he gave on quantum physics, Einstein once said to his students: "If you have understood me, then I haven't been clear." So if it sounds like nonsense to you as well, everything is fine. Nature doesn't take offence. It is there for us to discover, and that's it. But is it truly real?

Well, some people did take quantum tunnelling very seriously and tried to find practical applications for it. Amazingly, they succeeded.

About thirty years ago, working for IBM in Zurich, Switzerland, German physicist Gerd Binnig and Swiss physicist Heinrich Rohrer became convinced they could use quantum tunnelling to browse any surface at a phenomenally small scale, and see what the surface looked like. They believed quantum tunnelling could allow them to actually see the atoms.

Normally, an electron does not leave its atom if it has nowhere better to go. And normally, if there *is* somewhere else to go, it needs to be pretty close by, otherwise the electron can't get there. This is unless it uses its quantum power to tunnel through voids and jump over obstacles.

With an extremely thin and sharper-than-sharp needle connected to an electric-current detector, Binnig and Rohrer scanned the surface of a material, without touching it. Being rather far from it, they shouldn't have detected anything, the distance between the surface and the needle being too great for an electron to cross. But they did detect electric currents, corresponding to electron jumps.[*] And the closer to an atom on the material's surface the needle was, the more jumps they detected, the greater the electric current. They then mapped these currents on a chart and they got a 3D picture of the material, at an atomic level, with extraordinary details. They had built a microscope, a *scanning tunnelling microscope*, as it is now called, which could see the atoms themselves. Its precision is astounding: between 1 and 10 per cent of the diameter of a hydrogen atom. In other words, if hydrogen atoms had feet, a scanning tunnelling microscope would be able to count them, maybe even the number of toes.

Atoms of gold like the ones you found in your kitchen were scanned this way decades ago and scanning tunnelling microscopes are today used to picture how different types of atoms are intertwined within the matter that surrounds us, as well as in state-of-the-art human-made materials. With such a microscope, engineers have gained the ability to move individual atoms around. Quantum tunnelling is real. And it has practical applications.

For designing such a tool, Binnig and Rohrer were awarded the 1986 Nobel Prize in Physics.[†]

Electrons like the one you've tried to catch populate the outer reaches of all the atoms of the universe. And they are elusive. But despite being unable to describe exactly what they look like using

---

[*] In case you are wondering, virtual photons, the pearls of light that carry the electromagnetic force, do not carry any charge, so they can't be responsible for this.

[†] They shared the Nobel that year with Ernst Ruska, a German physicist who built another type of microscope called the electron microscope. 1986 was a magnifier's year.

the terminology of everyday life, scientists have learnt to accept their strange behaviour.

As far as today's science knows, electrons are not made of any smaller particles. Unlike atoms, they cannot be cut, or split, or broken at all. They are just made out of the electromagnetic field; they are an expression of it.

For being nothing but themselves, for being one of the most basic, fundamental expressions of the electromagnetic field, electrons are called *fundamental particles*.

By contrast, the evanescent pearls of light that earlier appeared in between the magnets and your fridge were called *virtual* particles. These were *force carriers*. They existed only to carry the electromagnetic force between electrically or magnetically charged particles.

Atoms, being made out of smaller constituents (like electrons and whatever makes up their core), are *not* fundamental particles. They are made out of many of them.

Now, electrons do not just interact with the rest of the world through virtual photons. They can also play with *real* photons, with the real light your eyes detect. This game of matter and light is what makes us see the world as we do.

As they are understood nowadays, real photons, like electrons, are also fundamental expressions of the electromagnetic field, made out of nothing else: they are pure ripples within an invisible sea, quantum ripples able to behave like waves *and* like particles.

A bunch of them are now washing over your hydrogen atom. They travelled a long way to get there. For about a million years they fought their way from the Sun's fusion core to its surface, which they reached about eight and a half minutes ago. Free, at last, to race through outer space unhindered by matter, they shot at the speed of light across the 93 million miles that separate our star's furious surface from our planet. Out of all the places they could have gone to, these photons ended up hitting Earth's atmosphere just a fraction of a second ago, only to charge through it and reach . . . your kitchen window. From there, there wasn't much left for them to do. They went through the windowpane and washed over your hydrogen atom.

Your mini-you watches them as they stampede through the kitchen, hoping to see them hit your atom. Instead, they all fly through it and smash against your kitchen wall.

Except one, which is gone.

Vanished.

Where did it go?

You look around, surprised, until you realize that your hydrogen's elusive electron is now wiggling differently than before. Considered as a wave enclosing the nucleus, its crests are closer to one another.

How can that be?

It became excited.

It swallowed the photon.

Remember that we first met this strange phenomenon back in Part Two when checking the first cosmological principle, some time ago.

But now something even more interesting is happening: after a short while, the electron suddenly spits out the *exact same* photon that disappeared, the photon it had swallowed, in a random direction.

After taking a moment to ponder this, you draw the only possible conclusion: that the most well-known fundamental particles of the electromagnetic field, namely the electrons and the photons, can and do interact with each other. That electrons and photons can turn into one another.

Mulling it over some more, you realize that you've actually

always known it: don't you feel warm when bathed in sunlight? Doesn't your skin heat up when, in wintertime, it faces a few burning logs in a hearth? Your skin, like all the matter in our world, is made of atoms whose outer layers are filled with electrons. When light from the Sun hits them, the atoms of your skin and their electrons catch some photons that are turned into excited electrons, electrons that wiggle somewhat faster, creating the heat your body enjoys (or not).

It is such an incredible discovery that I here say it again: matter and light can, and do, turn into one another.

Everything is a game of matter and light in this world of ours.

But not only that.

4 | *The Tough Electron World*

In the last two chapters, although you only saw a magnet interact with a fridge and skimmed the surface of an atom, you made great discoveries. You solved the mystery of electromagnetism's "action at a distance" and saw how matter and light can play with each other. Of course, this game is but one facet of our world, but it is a phenomenon that our humble human senses are built to be aware of. Light continuously hits our body, exciting electrons within our flesh, within our eyes and on our retinas, heating up the matter we are made of, giving it some energy. Atoms can also spit back the light their electrons swallowed, making us and objects "shine" with one or more colours, the colours of the atom – or the set of atoms – that swallowed them. That is what gives our eyes, our skin, our hair and clothes, all the plants and stones, a colour, just as it gives faraway stars a particular hue as well. Light rays hit a tomato; all of the visible light is absorbed to heat it up or to be stored within, except for red light rays, which, serving no purpose for the tomato's atoms, are spat back out, off on their journey to tell our eyes that we are looking at a lovely red tomato. Without the electrons and photons, we wouldn't see a tomato or each other, nor would we know what the rest of our universe is made of, or that it obeys the same physical laws far away as the ones that apply around us. But what is even more amazing is that, thanks to our senses, our bodies transform all these otherworldly interactions into sensations processed by our brains. Thanks to that, mankind figured out the science behind those interactions, and the existence of fields that fill in the entire universe. And this is not just amazing – it's plain miraculous.

Now what about that atomic core, the nucleus? Is it made out of electrons too? Is it yet another expression of the electromagnetic field? It has to be, in some way, for as far as you can tell, the whole hydrogen atom you are looking at is electrically neutral. The core must, then, have a charge too, opposite to that of the electron that surrounds it, so they both cancel each other out when seen from a distance. But how come you do not see it?

As your tiny self scrutinizes the hydrogen atom that floats in the middle of your kitchen, it suddenly occurs to you that this hydrogen fellow looks awfully like a lot of empty space compared to what it actually contains, whatever its core may be made of. That fact – the amount of emptiness lying in between the core and the electrons – is actually shared by all the known atoms in the universe.

Strange.

Why, then, doesn't a magnet just cross through the surface of a fridge, the vast, empty spaces of the magnet's atoms wafting past the vast, empty spaces of those of the metal door? Why does it instead stay stuck to it? Shouldn't colliding atoms just *not* collide, and cross like two clouds of vapour, without even noticing each other's presence? Well, no. Fortunately. Or the world wouldn't be solid. And the electrons are the reason why, not the nuclei. To figure out what that reason is, the atom of gold you've already prepared will come in handy.

The hydrogen atom you've looked at so far is the smallest atom there is. An atom of gold is bigger. You jump next to yours and look at it.

The first thing you notice is that it does not have one lone, wavy electron whizzing around its nucleus, but seventy-nine, all seventy-nine being identical to the lone, wavy electron whizzing around the hydrogen core.

The second thing you notice is that, however identical they may all be to each other, these wavy electrons do not share their territory at all. Ever. They simply and plainly avoid being at the same place at the same time, because it so happens that nature forbids them from doing so: whatever the atom they belong to, their wavy selves overlap nowhere, thus imposing very strict conditions on their potential cohabitation within any atom. They have no choice but to arrange themselves in layers, like an onion, around the nucleus, and that is exactly what they do. Only two electrons can fill the first, innermost layer. Only eight can settle in the second, eighteen in the third, thirty-two in the fourth and so on.

These numbers are known and they are the same for all the known atoms in the universe. What makes one atom different from another is linked to the number of electrons it contains, not to the nature of these electrons. Electrons are always identical.

Hydrogen, the smallest of atoms, has one electron whose orbital lies within the first electronic shell. Helium has two electrons. Their orbitals fill up the first shell. Neon, to take a third atom at random, has ten electrons. Its first two electronic shells are saturated. The chemical and mechanical properties of all the atoms are related to how full their external atomic shell is.

If you want to add an extra electron to an atom, you can't just put it anywhere you please, and certainly not within an already filled layer. Now, were the electrons dotlike particles, that would be hard to conceive. But although they can indeed be like little marbles under some special circumstances (you will hear more about that in Part Six), they also can *not* be, so as to behave instead as waves. And waves can very easily fill in some volume. And that is how, on a filled electron layer, there's no space left for a newcomer whatsoever. Were an extra electron (whether on its own or belonging to another atom) to really want to be part of an already built atom, it would either have to settle further away than the natives, where space might be available, or take the place of one that is already there, kicking it out. They just abhor having their wavy selves overlap. It's a dog-eat-dog world.

This non-cohabitation rule has a name. It is called the *Pauli exclusion principle*. It was discovered in 1925 by Swiss theoretical physicist Wolfgang Pauli,* who was awarded the 1945 Nobel Prize in Physics for it.

This exclusion principle is the reason why magnets stick to fridge doors without crossing them, or, perhaps more importantly, why you can't walk through walls, and why you don't fall through the floor. It also explains why you can hold this book in your hands: the atoms in its cover have outer electrons that absolutely refuse to give their place to the ones in your fingertips. And your electrons won't budge either. So they keep apart. And there's no way your own strength could force any of them to act otherwise. Electronic waves do not overlap. Ever. Don't try to run through a wall to prove me (or Pauli) wrong. Your nose would break long before the electrons noticed anything.

That being said, although electrons like their privacy, they do not mind being shared. And this allows them, rather fortunately for us, to build the matter we are made of, as you are now about to see.

You were about to plunge into your atom of gold but it will have to wait, for it so happens that an atom of oxygen is passing by.

You stare at it.

Smaller than gold, oxygen, with its eight electrons, is still much bigger than hydrogen.

Its first atomic shell is filled, but there is room for two more in

---

* It so happens that at the time, Pauli had just been dumped by his wife . . . for a chemist, something very hard to swallow for a theoretical physicist, and Pauli began to drown his sorrows in alcohol. No wonder his principle bears the name "exclusion." Still, ironically, from the depth of depression, he found the reason why we can live on the surface of our world without falling through it, even though he seemed to have lost such a reason for himself.

its outermost shell, the second one, which has six electrons and could contain eight.

The hydrogen atoms' lone electrons are not about to let such an opportunity pass.

And there are two hydrogen atoms nearby, so as soon as oxygen passes by, *hop*, the first hydrogen's lone electron jumps over and settles down within the oxygen's family, never to be alone again.

And *hop*, just as you see it happening, the other hydrogen's electron fills in the last spot.

And since all the electrons in the universe are exactly identical, no one can tell who was there in the first place and who arrived later. The perfect assimilation.

Bound to their electrons by virtual pearls of light, the nuclei here have no choice but to follow them, and so the three atoms are now stuck to one another. Two hydrogens and one oxygen are forced to cohabit.

This done, there's no more space available for any extra electron. The whole construction is stable.

By sharing their electrons as above, atoms become part of larger structures, which are called *molecules*. The molecule you've just seen being built is made of two atoms of hydrogen and one of oxygen.

Two Hs and one O.

H_2O.

It is water: the most precious molecule for life as we know it.

On a universal scale, water is not usually assembled in your kitchen, though, but rather in outer space, inside the huge clouds of stardust that are scattered within galaxies, and which astronomers call *nebulae*.

Within these nebulae, the oxygen forged in exploded stars mixes with hydrogen, which can be found everywhere.

When stars die, they send their seeds away, paving the way for water molecules to be built. And many other molecules as well.

By sharing one or more of their electrons, many atoms can be bound together, in many different ways, to form chains of various complexity. Nature has built molecules of different sizes and properties this way, from rather tiny ones (water molecules are made of three atoms only) to extraordinarily long ones, such as your own DNA, which, with its billions of joined atoms, carries all the information needed to build someone like you.

It is to shed some light on the genesis of these molecules that started life on Earth, and to unravel the mystery of the origin of the water that today covers 70 per cent of our planet's surface that, within the last decade, many satellites have been sent to outer space. Does our water come from asteroids that struck our planet some 4 billion years ago? Or from comets that did the same? And did these rocks and iceballs carry some, or all, of the molecular seeds of life? We should soon know, for many of these satellites are now on site, or on their way there.

In the meantime, we do know one thing, however: only six atoms were necessary to build all the molecules required for life to thrive on Earth: Carbon, Hydrogen, Nitrogen, Oxygen, Phosphorus and Sulphur. The so-called CHNOPS.

Incidentally, since your whole body is made out of molecules that are made out of these atoms, assembled in various ways, you are a CHNOPS. No offence meant at all.

Now, as you contemplate your chnopsy body, another question pops up in your mind: since both you and the air are made out of these atoms sharing their electrons, how is it, then, that you (very fortunately) can walk through air but can't walk through a wall?

An important question indeed.

As far as we know, air is filled with atoms, which have as many electrons as one could want, so they shouldn't let you pass. No. That's Pauli's rule.

The answer is that the atoms in the air are not all sharing their electrons and hence don't hold on to each other that much whereas, forming a solid, yours are. Instead of stopping you from moving at all, the electrons that surround the atoms that make up air move away as yours force their way through, incidentally bumping into each other, creating some wind. That, incidentally again, is the difference between a gas and a solid.

In liquids, nearby atoms are a little more tightly bound to each other, but not enough to stop you, unless you try and enter too fast, like when diving off a cliff into the iron-grey sea. In solids, atoms don't move aside unless you force them hard to do so – think of sharp scissors cutting through paper.

Now, instead of fighting for its position, an electron can also be forced to leave, offering an empty spot for another electron to fill. When an atom loses an electron (after being hit by a powerful photon of sunlight, for instance), the combined charges of the core and the electron(s) do not add up to zero any more. Atoms stripped of one or more electrons become what scientists call *ions*.* Ions tend to look for something to bond to, to form a molecule. In fact, they're desperate to find electrons. In the terminology of physics, they are fiercely *reactive*.

Conversely, bonds created by electrons within a molecule can also be broken. Energy is usually released during such a process, and that is what eating food is good for. Chemical reactions inside your body break up the molecules contained within the food,

---

* Atoms that have somehow gained one or more electrons are also called ions. Ions are atoms that do not have their natural amount of electrons.

releasing energy that is then used in many ways by your organism, to keep you alive.

All right.

This wraps up our survey of the electron's tiny world. You've only skimmed the outer part of three atoms and yet you've already figured out how modern science understands pretty much everything our bodies experience on a daily basis. So, before heading for the still mysterious atomic core, I shall summarize for you what you've been through over the last couple of chapters.

The outer parts of all the atoms throughout the universe are smeared-out, wavy, massive electric charges called electrons. They are fundamental particles of the electromagnetic field and they are very protective of their personal space. The Pauli exclusion principle forbids any two electrons from being at the same place in space and time. Even though there is more void than anything else in all the atoms of the universe, that is the reason why you cannot walk through a wall, do not fall through a chair, or through a bed, or through pretty much anything solid. Living would be tricky otherwise.

Pauli's rule also brings about the structural and chemical differences between different atoms: since electrons cannot all crowd as close to the nucleus as is possible, they occupy a series of onion-like layers around the atomic core, filling available spaces only, making the atoms grow with the number of electrons they contain.

Electrons, it has to be said, are not the only particles subjected to Pauli's exclusion principle. Other particles are, too – but not all. Light, for instance, begs to differ. You can pile up as many photons as you want in as small a place as you want. They won't mind. In fact, they quite like it, and the more similar two photons are, the more they tend to cuddle each other, like penguins in the cold.

Lasers are a consequence of such fondness: they are highly concentrated, highly energetic beams of identical photons.

Now, having got this far, you might be left with the impression that electrons and light are the only particles that count in our universe. But that is not true. You will soon see that there are others within atomic nuclei, but I just want to stress that there are even particles around us that care neither about the electrons' wish for privacy, nor about their existence at all. Or about anything else we know, for that matter. They are particles that do not belong to atoms. Some of them are actually so aloof that most of the time they shoot through anything and everything, leaving barely a trace of their passing. For these tiny particles, the universe must seem pretty dull and empty. Even Earth. Even you. You'll meet them soon enough.

For now, however, you should celebrate again! With what you've just learnt about electrons and light, you know what only a handful of people knew half a century ago, and most of them were rather bright, since they got a Nobel Prize for figuring it out.

But there is more.

Thanks to them, you can now explain pretty much everything that happens around you, from the colour of a tomato to the solidity of a wall or ground to the reason why magnets jump from your fingers to stick to fridges' doors.

Everything you and I and all our friends experience on a daily basis is regulated by matter and light playing with each other, turning into one another, and by electrons categorically refusing to share their bit of spacetime with a copy of themselves.

The next time you embrace someone, feel free to imagine virtual pearls of light being created and becoming frantic as you get closer and closer to each other, before the electrons abide by Pauli's rule and decide that no, you cannot get any closer. I'm not sure you should talk about this amazing fact on a first date, but I leave that decision to you.

Before you continue your journey through the matter we know, here is another good piece of news: in 2014, experiments done in the impressive underground scientific laboratories of the European Centre for Nuclear Research (CERN) underneath the French–Swiss border confirmed that humanity had theoretically discovered everything there is to know about the matter we are made of.

Everything.

This does not mean there are no mysteries left (you will see plenty of them in Part Six). But it does mean that, since 2014, we have a picture of our universe's known contents that corresponds to pretty much everything we could possibly probe or find within the range of modern technology.

That picture includes atomic nuclei, the atomic cores you are now ready to go examine.

And if you somehow have a hunch that you will again find weird things down there, you are absolutely right.

5 | *A Peculiar Jail*

Your coffee is getting colder and colder and your arm, holding the milk, aches. But you don't care.

Your mini-you just decided to dive deeper and deeper within one of the hydrogen atoms that formed the water molecule right in front of your eyes, towards its core. Many evanescent pearls of light (the virtual photons you saw between your magnet and your fridge) are appearing and disappearing all around you, confirming that the core you are aiming for is electrically charged, destroying the idea that there is only nothingness between an atom's electrons and its core.

Still, compared to the size you guessed the atom to have, you cross immense distances before reaching the hydrogen core.

But you eventually find it.

Like the electron spinning around it, the core of the hydrogen atom does not seem to have a particular shape, yet it does have a mass. But it is heavier. Much heavier than the electron: 1,836 times so. And it does have a charge, indeed the exact opposite of the electron's charge.

It is called a *proton.*

It is bigger than the electron but compared to the size of the atom itself (that's the volume spanned by the electron), it is extraordinarily tiny. New Zealand–born British physicist Ernest Rutherford discovered its existence in 1911, three years after being awarded a Nobel Prize in Chemistry for his work on a then very new phenomenon called radioactivity. What he did not know, however, what he *could* not know, is that unlike the electron, the proton is not a fundamental particle. It has a world within.

Not to waste time attempting the impossible, you close your eyes and spread your arms to *feel*, yogi-style, what that proton's inner world is about.

Immediately overwhelmed by a force so strong that anything you've experienced so far feels like child's play, you open your eyes again straight away.

Electromagnetism can easily overpower you: some magnets are so firmly stuck to each other you'd never be able to move one away from the other.

Gravitation can also overpower you, and it actually does: you'll never be able to jump free of Earth's gravity.

But this is another level of power altogether.

Within the proton, within what looked like a fuzzy, cloudy sphere, you glimpsed countless virtual particles appearing and disappearing, like the electromagnetic pearls of light you saw between the magnet and the fridge, or between the electron and the proton. But these are not virtual photons. They are the carriers of a new force, and that force, together with the quantum field it belongs to, is the one that stabilizes all the matter in the universe.

Without it, everything we know would disappear in a snap of the fingers. Everything. Your body included.

The virtual particles that carry this amazing force, the force that keeps matter intact, are hundreds of times more powerful than the photons carrying the electromagnetic force. They are the force carriers of the so-called *strong interactions*.

But if these were "just" the force carriers, why didn't you see the fundamental particles of that new field? Virtual photons made charged particles interact, so what is it that interacts here?

Without any second thoughts, you jump inside the proton, close your mini-eyes again, raise your mini-hands, and probe . . . feel . . . search for the purpose of carriers of such a strong force . . . Surrounded by so much energy, a huge effort of concentration is necessary, but eventually you manage it. You can make out three things, three fuzzy, wavy, heavy little things that scientists have called *quarks*. The name may sound strange, but don't all names before we become used to them?

No one, besides you right now, has ever actually seen a quark on its own. They do not even exist on their own – the strong, virtual little fellows that continuously appear and disappear around them just won't let that happen. The further apart the quarks are, the more ferocious the strong force carriers become, bringing them back close to one another much more efficiently than any other known force in nature.

For the three quarks that live inside the proton, life is therefore rather confined. Jail-like, really.

And their virtual prison guards, the strong force carriers? Who are they? What are they? They are not photons, that is for sure. They are not part of the electromagnetic field, remember – they are expressions of another field entirely, the *strong interaction quantum field*.

And they are so efficient at their job of gluing the quarks together that they have been called the *gluons*.

Quarks and gluons.

They make up all the protons of our universe.

Now here is something strange about this tiniest of jails your mini-self is visiting: most of us definitely believe that if you find yourself behind bars, as a person, then freedom means being as far away from your cell and its guards as possible. Well, for the quarks held in the protons, felons or not, it is the other way round. For them, freedom lies in short distances. The closer they get to one another, the freer they get to do whatever they want. Quark freedom is a very strange concept indeed: a world of possibilities opens up for them as soon as they get closer to one another.

For having discovered this peculiar type of freedom, three

American scientists, David Gross, Frank Wilczek and David Politzer, received the 2004 Nobel Prize in Physics. A hard concept indeed. So hard that when I met David Gross and Frank Wilczek in Cambridge, a couple of years before they received the prize, I remember wondering whether I should ask them to pay me back the money I spent on headache tablets trying to understand their work.

Quarks and gluons.

Elementary quarks, made out of nothing but themselves.

And gluons.

The carriers of the strongest force we know, the *strong nuclear force*, which keeps quarks confined, only allowing them to be free when they are holding close to one another, thus guaranteeing that the matter we are made of does not split apart.

Quarks and gluons.

Strange names indeed, names used to describe the essence of a reality so far from our daily lives that it may well sound rather insignificant. The strong force, however, with its quarks and gluons, involves about 99.97 per cent of the mass that makes up our bodies. Were a 130-pound person to lose all his or her quarks and their binding gluons right now, he or she would instantaneously slim down to weighing less than an ounce. And be dead, obviously.

To understand what mankind has so far discovered about our reality, to even figure out what our reality is made of, quarks and gluons are therefore very necessary. And that, it seems to me, is a fairly good reason to study them. Notwithstanding the fact that they will shortly allow us to travel back to about a second after the birth of space and time.

Now, as we have already said, the field these new fellows belong to is called the strong interaction field, or strong field. It is a quantum field, of course, so most of the weird quantum behaviour found earlier, involving electrons and light – disappearing and appearing elsewhere, or "tunnelling," for instance – does apply here as well. But what is important to underline here is that the strong field is *not* the same as the electromagnetic field, and yet it also fills the entire universe. It is another sea, if you will, whose drops are quarks and gluons rather than electrons and photons. And nothing prevents particles from belonging to both: being electrically charged, the quarks belong to the electromagnetic field as well as to the strong interaction one. They can interact with the force carriers of either: through light *and* gluons. But over short distances, the gluons are much, much more powerful than light.

Now what about this new sea? What are its fundamental particles?

The strong field has six of them, six different quarks which can pop out of the strong field anytime, anywhere, if sufficient energy is present. Only two of them are found within an atom's core, though. They are the so-called *up* and *down* quarks. There are two up quarks and one down in every proton in the universe, so it is fair to say that protons have more ups than downs, which perhaps explains why they are happy in their subatomic jail.

But protons are not the only quark jails that exist, as you will now see within your atom of gold.

Bored with the hydrogen, your mini-you jumps back to above the kitchen table where you cut your treasure to bits.

Your atom of gold is still there – you dive into it.

Buried deep beneath the seventy-nine electrons that swirl around it, its nucleus is much, much bigger than the hydrogen atom's. To match the charge of its seventy-nine electrons, you find seventy-nine protons. But there are also other fuzzy spheres surrounding – separating? – these protons. Uncharged spheres. You can count 118 of them.

Being electrically neutral, they are called *neutrons*. They, too, are quark jails and they were discovered by the English physicist Sir James Chadwick, who happened to be a deputy of the extraordinary Rutherford.* Chadwick received the 1935 Nobel Prize in Physics for his discovery.

Within every proton, gluons confine two up quarks and one down. Ups have the majority. Within neutrons, it is the other way round: downs lead two to one.

Now, how do all these jails stack up to make an atomic nucleus? Why don't they move apart from each other? Or collapse? The protons are all positively charged, after all. They should repel each other.

But they don't. Why? Because the strong field and its force carriers prevent them from doing so, albeit in a very strange way. A residual way.

To figure out what this means, your mini-you boldly decides to closely watch the elusive gluons that guard the quarks within a proton. They are there. You can't really see them, but you can feel them, in yogi mode. They appear and disappear to keep the quarks from wandering away on their own.

But suddenly, something very strange happens.

Something has left. Something has jumped out of the proton. But what was it? A gluon? Why not, after all? They are guardians, not prisoners . . .

But no, it wasn't one gluon.

Not one on its own, anyway.

You sharpen your yogi sense . . . and there you go.

---

* Rutherford, one of the most impressive experimentalists of all time, had also discovered that atoms have a nucleus (I mentioned him earlier on in this part). Chadwick was working at the Cavendish Laboratory of the University of Cambridge, England, which Rutherford directed.

The gluons, it so happens, don't leave on their own at all. They have to find another gluon to pair with. A friend. Pairing with the right one, they turn into something else . . .

You look around and right there, to your left, in between two quarks, it is happening again.

A gluon pops out of the background field and so does a friend of his, another gluon, and they now stick to each other and . . . *pop!* Just as light can be transformed into an electron, these two gluons transform themselves into two quarks! A quark duet that is not bound to the other quarks by the gluons any more! They become free, as a new entity, to leave the quark jail they belonged to!

You watch them leave.

They are heading straight towards a nearby quark jail. They have in fact become the carrier of yet another force, a force that acts not upon quarks, but upon the quark jails themselves. As they reach one, they transform back into gluons and start guarding the quarks there . . .

It is thanks to such exchanges that neutrons and protons coexist within atomic cores. By travelling from one jail to another, two gluons-turned-quarks make sure that atomic nuclei remain stable. The exchanged particle, the quark duet that travels between jails, is called a *meson*. And the force it carries is called the *strong nuclear force*. It is an attractive force. And a very strong one.

For having predicted the existence of mesons long before they were found in experiments, Japanese theoretical physicist Hideki Yukawa was awarded the 1949 Nobel Prize in Physics.

In a funny twist, the boiling soup of quarks and gluons that can be found inside all protons and neutrons happens to also be responsible for the missing mass we mentioned a long time ago, the missing mass that makes stars shine.*

---

* If you've forgotten about this, it was here: Part One, chapter 3, page 20.

Within stars, as you now well know, small atoms are fused together to build new, bigger ones. This means that stars fuse neutrons and protons together, and it so happens that once fused, these neutrons and protons do not need as many virtual gluons to guard their quarks (or mesons to guard their jails) as they did when on their own. It is a bit like when two companies fuse: some people are made redundant, and they are fired . . . In star cores, the redundant gluons and quarks and mesons are also fired. Because they carry some energy and because energy is mass, firing them lowers the mass of the newly fused core. That is why all cores built by fusion are less heavy than the fused ones taken separately. As opposed to *people* who have been fired, however, this missing mass is turned into energy, the exchange rate being given by $E = mc^2$, making stars shine.

Deep inside stars, gravitational energy is thus used to forge atoms, a process that also involves mass being transformed into light and heat and many other particles that are around but which our eyes do not see. Although most of our reality is hidden to our senses, everything is linked in this universe of ours.

6 | *The Last Force*

So far, you've become aware of the existence of two quantum fields, namely the one responsible for all electromagnetic interactions and the one that gives rise to the strongest force known to mankind, the rather appropriately named strong interaction, with its residual strong nuclear force.

In a sense, these forces, and their fields, are powers of construction. Even though magnets can either attract or repel each other, the electromagnetic force makes sure that the electrons stay around atomic cores. The electrons could move away, or collapse on the nucleus. They don't. Virtual pearls of light prevent them from doing so. The electromagnetic field gives the atoms their electronic stability and ways to share their charges, to build up molecules, to build up the matter we are made of.

The strong nuclear force, on the other hand, is in the care of the atomic cores themselves. It keeps neutrons and protons together, building atomic nuclei. Without it, all nuclei would break apart and we would all instantaneously turn into a mist of protons and neutrons. And so would Earth and everything else.

And to root all this, the strong interaction keeps the quarks confined within these protons and neutrons, binding them with gluons popping out of the background.

So, you've travelled inside these two fields, you've seen that their interacting particles and force carriers give the world its hard, although elusive, tangibility. You've seen photons and electrons play with each other, and turn into one another. You've seen gluons and quarks all wiggling within the cores of precious gold atoms and those of common hydrogen, the smallest and most abundant brick of matter in the universe, the one that stars fuse in their hearts to forge the substance you and I are made of.

Hydrogen, whose exhaustion sooner or later triggers the death of all the stars in the universe . . .

Pondering this last remark, you suddenly remember what will happen to our Sun, 5 billion years from now, and you immediately recover your normal size, leaving your mini-you floating somewhere else, in a world too small for your normal eyes to see.

Your perception of the universe has changed hugely since you were lazily looking at the stars from the comfort of your tropical-island beach. You now know that nothing is empty, that everything interacts with everything, all the way to the innermost parts of the atoms which were built out of, and remain whole thanks to, extraordinarily remote interactions.

•

The sky outside your kitchen window is now turning red. The Sun is setting somewhere to the west, illuminating the flat base of the clouds with fiery colours.

Your arm aches for having held the milk for so long, but as you now sip your (cold) cup of light coffee, you mindlessly walk a couple of steps to stand next to the window and stare outside at the sky, and you suddenly better understand what it means to be part of a star's family.

All the stars throughout the universe radiate and shower their surroundings with light and particles, all direct or indirect by-products of their atomic-nuclei-fusion power-plant of a heart. And whereas their gravity, the curve they create in spacetime, makes every single close-by or passing object fall towards them, these winds of particles and light are blowing outward, spaceward, towards the faraway, radiating ripples of invisible background fields that fill in everything.

The universe really is like a vast ocean, and it so happens that some (very serious) space engineers have imagined building spaceships with huge sails to capture these solar winds and propel their ships towards the outer universe, like cosmic sailors riding the slopes of spacetime without needing any fuel . . .

Night has now settled down and you haven't moved. The sky has cleared. You are staring at the stars. There aren't many, though, light pollution is far too bright. Still, you now know the stars you would otherwise see from here are not the same as the ones you saw from your tropical island. You are now collecting photons emitted by stars dwelling in a different part of the Milky Way. But they are stars all right, huge balls whose gravitational energy is building large atoms out of small ones, merging their cores together.

Quite remarkably, and rather contrary to what we humans are used to, everything seems to be a force of construction out there.

Seems, yes, for you haven't yet seen everything that is known.

For that, a third quantum field is needed.

A third sea that fills in the entire universe just like the other two, a sea whose fundamental force carriers are neither photons nor gluons nor mesons.

And this one can somehow be viewed as a field of destruction, a field that undoes what the others have done. It is the last of the four forces that rule our universe.

This last force is also a nuclear force: just like the strong one you've just discovered, it only acts upon the constituents of atomic cores. But this one is much weaker than the strong force and it has therefore been called the *weak nuclear force*. The ubiquitous quantum field out of which its fundamental particles and force carriers are built is called the *weak nuclear quantum field*. The spontaneous splitting of atomic cores, a process known as *radioactivity*, is one of its attributes.

Now before you witness radioactivity in action, it may be worthwhile for you to recall that radioactivity has claimed the lives of many of its discoverers. Unaware that lethal invisible lights that slowly shattered their bodies were irradiating them, they processed raw, highly radioactive materials with their bare hands . . . The amazing Polish-born French scientist Marie Curie, the only person ever to be awarded a Nobel Prize in Physics (in 1903, for co-discovering radioactivity) *and* a Nobel Prize in Chemistry (in 1911, for discovering two new atoms: radium and polonium), was one of them. She may not have known why she died, but what you are about to witness is what she would have seen had she the knowledge we have today, together with the handy ability to turn herself into a mini-Marie.

As you empty your cold coffee into your sink, your mind shrinks back into your mini-self and your mini-eyes take a moment to adapt to the darkness.

You are back near your atom of gold.

It is right in front of you, an atom so strong and solid that an energy greater than that of a star's gravity is needed to forge it. Gold is not built during a star's life, but during its explosive death. When our Sun dies, it too will create some gold, which, who knows, may one day be paraded on the finger (tentacle?) of a future alien species.

As you look at it, however, this atom of gold doesn't look nearly as valuable as pretty much all of mankind seems to believe.

Why is it so coveted, then?

Does it change as time passes? Does it catch passing atoms to build extraordinary molecules?

You wait a bit, to see if it is so.

But no.

Nothing happens.

And there you go.

The fact that nothing ever happens to it is one of the reasons gold is so valuable. Gold does not rust. It does not oxidize (that's what happens when some electrons from an atom of oxygen bind to it). It does not corrode. And when you have a nice lump of it, it is the most ductile of all metals: you can draw gold out into the longest, thinnest wire of them all (platinum and silver would break long before). Put many gold atoms together and you can also easily mould the lot into pretty much any shape you like. And whatever you do to it, it will still conduct electricity, meaning that an electron introduced on one side of a long chain of gold atoms will wave its way along the chain and exit it from the other side.

All these exceptional properties can lead to practical applications that may not always be very apparent in a wedding ring, but which are priceless.

Add to this the fact that gold is rare, hard to mine and forged by the death of a star, and one understands why it is expensive. We'll leave it there, however, because nothing is happening to it at all, really.

To see something different, you'd need another atom, and funnily enough, one is dropping by.

And it is bigger.

As far as you can tell, it has ninety-four electrons swirling around a core made out of ninety-four protons and one hundred and forty-five neutrons. Two hundred and thirty-nine quark jails. That's forty-two more than gold.

This atom is one form of the infamous element called *plutonium*. And because it has 239 quark jails, it is called plutonium-239. There are other types of plutonium, just like there are other types of gold* besides the one you've found in your kitchen. These can have more,

---

* Or hydrogen or any other atom, really.

or fewer, neutrons in their cores, but they always have the exact same number of protons, or they would not be plutonium – or gold – any more.

And whereas gold wasn't very interesting to watch, something tells you that a strange phenomenon is about to happen, spontaneously, inside the nucleus of plutonium-239.

Without any hesitation, you travel through layer after layer of its electron shells. You traverse huge voids filled with virtual photons. And then there is the nucleus. The 239 quark jails are right in front of you. The strong nuclear force is keeping them neatly stacked, but your gut tells you to aim for one of the neutrons.

You plunge right in.

There are two down quarks and one up quark in there, tightly kept together by the sturdy gluons.

Just as you settle down, however, one of the down quarks is hit by a virtual particle you have not seen before, a virtual particle that spontaneously appeared only to transform your down quark into an up one. The neutron it belonged to thus just turned into a proton, creating havoc. Now the whole atomic core is out of equilibrium. The effect is instantaneous, and dramatic.

A sixth sense tells you to run for cover and your mini-you rushes out of the core and the electron layers only to see the plutonium nucleus split and split and split again into other, smaller ones, all of them trying – but sometimes failing – to take some electrons with them. Extremely energetic particles are fired off at each stage of the process, including yet another one which you have never seen before. Your plutonium is decaying. Right in front of you. And all the products of this decay are now shooting away. A firework that eventually burns itself out. Unless many other plutonium-239 atoms are around. But there aren't, in your kitchen. So everything quickly quietens down.

You just witnessed one aspect of the fourth known force of nature: the weak nuclear force, with its virtual force carriers that can turn quarks into one another. These force carriers are called the *W* and the *Z bosons*.

What you have just seen is the decay of an atom into smaller, more stable ones. It was the spontaneous fission of an atomic core, the very opposite of its fusion. A *radioactive decay*. This is what *radioactivity* is all about, and the weak nuclear force is in charge of it, with its W and Z bosons to carry it.

Wolfgang Pauli – the same Pauli who came up with the exclusion principle – studied such atomic decay about a hundred years ago. He didn't know about fields, as you now do, but when comparing what he observed before and after a radioactive decay, he realized that there was some energy missing. So he forecast the existence of a hitherto unknown particle that could be blamed for carrying that energy away, a particle with minute mass, a particle carrying no electric charge whatsoever, a particle so elusive that, once it has been fired, it shoots through all the matter we know almost unhindered.

This new particle is now known to exist. You just saw it. Out of all the particles fired out by the radioactive decay, it is the one you had not seen before. It is called a *neutrino*.

US physicist Frederick Reines and his colleagues detected it experimentally in 1956 and Reines received the Nobel Prize in Chemistry for it almost forty years later, in 1995. As he once said, neutrinos are the tiniest quantity of reality ever imagined by a human being. Today, we know that these neutrinos (there are many of them) are only subjected to the weak nuclear field and gravity. They are entirely unresponsive to the electromagnetic and strong fields.

For them, atoms really are what they looked like to you in the first place: empty.

And that is a good thing.

Why?

Because were the neutrinos to interact with atoms, we would be in trouble, for they are abundantly produced within the Sun.

Very abundantly, in fact.

About 60 billion neutrinos shoot through every square centimetre of your skin.

Every second.

And they don't even notice you. Not a single one of them.

However vexing it may sound, they can't tell the difference between you and, say, nothing. They shoot through you. And then through the Earth.* And they continue their journey spaceward, as if neither you nor our planet had ever been there in the first place.

•

Now, we've all been taught that radioactivity is dangerous, and that one should, as much as possible, run away from radioactive materials like plutonium or uranium or radium or polonium – and rightly so. But since the neutrinos can't really tell the difference between you and nothing, they can't be the reason for this danger.

The reason has to do with the other particles fired during a radioactive decay, and luckily enough, you are already familiar with them.

When the core of an atom decays, it splits and can emit neutrinos and quark jails and electrons and light. And the last three are dangerous.

The largest of these consists of four quark jails bound to each other: two neutrons and two protons stuck into a lump. It is called an *alpha particle* and actually corresponds to an atom of helium that

---

* That is during the day. At night, they still shoot through you, but *after* having passed through the Earth.

has been stripped of its electrons. To become an atom, that core therefore needs to steal two electrons from somewhere else, a feat it can achieve in several ways. It can strip some out of nearby atoms (rude), it can share some with nearby atoms (altruistic) or it can adopt wandering ones (Samaritan).

In the first case, the atom stripped from its electrons starts to look for other electrons itself . . . When there are living creatures nearby (like us, like you in your kitchen), weird chemistry can happen with electrons being stolen from atoms on the skin, leading to what is called radioactive burns. That's why alpha particles are dangerous.

The second type of particle that can be fired by an instance of radioactive decay is a very energetic electron, which can knock off other electrons far away (leading to the same type of danger), while the third type is a very energetic photon, a *gamma ray* – we met them on our earlier journey through the cosmos, remarking on their incredibly high, energetic frequency.

A gamma ray can, just by hitting an atom, strip it of one of its electrons, turning that atom into an ion eager to find another electron, creating burns on our skin again.

But gamma rays can also do much worse.

Nothing obliges them to stop at the surface of our body. They can penetrate it and cause local havoc deep inside, not just by kicking electrons out of their atomic home, but also by breaking molecules, like DNA molecules, in the heart of our cells, thereby changing the instructions used by our organisms to create everything our bodies need to live. The outcome usually is cancer and/or genetic mutations.

All these potential outcomes are scary. There's no way one could argue otherwise. But there is a bright side too: just like gravity and electromagnetism and the strong interaction, radioactivity, despite being a force of destruction, is a natural process that occurs all the time, everywhere, even within your body, at a very slow rate. Only if one is exposed to high levels of radiation should one worry.

In fact, everyone should be very grateful radioactivity exists at all. It can kill you, yes, but without it you wouldn't have been born in the first place. On Earth, deep under your feet, our planet happens to contain many atoms that do decay, all the time. Less so now than in the past, but still, Earth's mantle *is* radioactive. When atoms decay there, the particles they emit bump into their neighbours and generate heat, the very heat that contributes to keeping our planet warm. Without radioactivity, there would be no seismic or volcanic activity. The surface of the Earth would have been dead cold billions of years ago. Life as we know it would probably not exist at all.

Radioactivity breaks the atoms. Radioactivity kills. But it is needed to heat up our world, giving us back some of the energy that stars stored within the atoms that built our home planet.

Now a last little comment before letting you embark on a journey towards the origins of space and time: atomic energy as a whole, through fission or fusion of atomic cores, involves extreme energies, and these energies are what mankind tries to harvest, with more or less efficiency, within nuclear power-plants. We can only hope that such technologies will one day become clean and safe, for their potentials are astonishing.

Despite their having rather bad press, and despite their past unjustifiable use, we should never forget that without nuclear forces, we wouldn't exist. Without radioactivity, life on Earth would be impossible.

Life as we know it, that is, of course.

Part Five

To the Origin of Space and Time

1 | *To Be Confident*

When I started to be interested in what some might call hardcore theoretical physics, I was about twenty-two. I had studied pure mathematics for some years before and was very fond of its beauty. As the Greek philosopher Plato said – about twenty-five centuries ago, when no one had a clue what the heavens were about – mathematics is the language in which gods speak to people.

When my application to study advanced mathematics and theoretical physics at Cambridge University, England, was accepted, I immediately thought: Great! Time for some deep thinking about the real world!

Little did I know what was about to happen to me, just as you possibly don't have a clue what is about to happen to you in the coming chapters.

During the summer that preceded my first year at Cambridge, I read a few textbooks as well as works by the masters of the past and present, to get a clearer feeling for what science might have to say about the world around us. I concentrated especially on the quantum world. After all, as we found out in Part Four, the world of the very small lies at the root of everything we are. It is there that we find the building blocks of all that our universe contains – indeed to even use Einstein's general theory of relativity one needs an understanding of what our universe contains, or its equations won't tell us what our universe looks like at large scales.

Many Nobel Prizes in Physics have been awarded to scientists for breakthroughs regarding the very small.

Needless to say, I was very excited about the journey I was about to go on, and as I began to get to grips with the theories of these intellectual pioneers, I started to write down some of their incredible thoughts, to make sure I was getting it right:

I think I can safely say that nobody understands quantum mechanics.

—*Richard Feynman, Nobel Prize laureate for Physics in 1965*

The Lord God is subtle, but he is not malicious.

—*Albert Einstein, Nobel Prize laureate for Physics in 1921*

No language which lends itself to visualizability can describe quantum jumps.

—*Max Born, Nobel Prize laureate for Physics in 1954*

Those who are not shocked when they first come across quantum theory cannot possibly have understood it.

—*Niels Bohr, Nobel Prize laureate for Physics in 1922*

I have second thoughts. Maybe God is malicious.

—*Albert Einstein*

Such statements, from the founding fathers of the field, would be enough to shake the belief of even the most confident of students. Still, alongside two hundred other young men and women from around the globe, I sat through mind-boggling lectures and passed what was at the time called the Part III Exam of the Mathematical Tripos, arguably the oldest maths exam in the world. It still consisted mostly of pure mathematics, and the amount of new material we learnt was so great that we had little time to really think about the philosophy of it all.

And then came the plunge.

Nine months after my arrival at Cambridge, Professor Stephen Hawking, one of the most famous (and brilliant) physicists of our time, offered me the chance to become his PhD graduate student, to work on black holes and the origins of our universe. Deep thinking was about to become compulsory. So I spent the following summer having another look at everything I could find out about, well, everything – and I reached pretty much the point you've arrived at now in the book. With Hawking as a supervisor, I was about to put it all together and to reach much, much further. Now it's your turn to do the same.

What is there left to see?

Well, here is a teaser.

In 1979, a very special Nobel Prize in Physics was awarded to three theoretical scientists: Sheldon Lee Glashow from the US, Abdus Salam from Pakistan and Steven Weinberg from the US.

For years, scientists had been trying to understand some rather peculiar aspects of the weak nuclear force that you recently saw in action. And Glashow, Salam and Weinberg discovered something incredible: that electromagnetism and the weak force are but two aspects of another force, another field, that existed a long time ago. They found that during the early days of our universe, at least two of the invisible quantum seas that fill our reality were once just one, the so-called *electroweak field*.

This was an extraordinary breakthrough in its own right (hence the Nobel Prize), but it also paved the way for something much, much bigger: the tantalizing prospect of unifying *all* the known forces of nature into just one force (and therefore one theory).

The quest for such a unification lies behind everything you'll experience between now and the end of this book. With this goal in mind, you will travel towards the origin of space and time, within a black hole, and even outside our universe.

In order to get there, however, you will first need to figure out what is left when one empties a place of everything it contains.

You are still in your kitchen.

The night is dark, and quiet.

If you thought that the world was beautiful before, it is now utterly transformed by what you have learnt on your travels. Everything seems somehow deeper, charged with power and mystery.

Even your humble kitchen.

The air around you is filled with floating atoms, sliding down the Earth's spacetime curve.

Atoms first assembled in the cores of long-dead stars.

Atoms within you, everywhere, disintegrating in radioactive decays.

Beneath your feet, the floor – whose electrons refuse to let yours pass, thus making you able to stand and walk and run.

Earth, your planet, a lump of matter made out of the three quantum fields known to mankind, held together by gravity, the so-called fourth force (even though it isn't a force), floating within and through spacetime.

This all sounds so absurd, or plain miraculous, that you decide to brew yourself some more coffee and make your way to the living room and sit down on your cosy, solid, reassuring old sofa.

You try to bring some order to all these thoughts that are bumping around in your mind. Is the meaning of life hidden somewhere out there, beyond what we have already seen together? And indeed does what you've learnt so far actually make any sense?

•

Before you head off to places even more remote than those you've seen so far, let me say this: unravelling the mysteries of the world is a work in progress. Science may not have all the answers, although it has many. It depends what your expectations are, in truth, for I must warn you now that the end may not make any more sense than the beginning. As US theoretical physicist Edward Witten once said, away from the safety of your home, the universe was not made for your convenience.*

It is probably worth keeping that in mind as we cast off into altogether darker seas, because however humbling such a statement may be, it offers us all the extraordinary freedom to interpret what we see in a personal way. And that is a good thing. For the more different points of view there are, the better for humanity, and the better for science.

Now, as I hinted at the end of the last chapter, before we push confidently through the doors of the unknown, we first need to become comfortable with a concept that scientists have called the *vacuum*. It is the basis of how our quantum reality is currently understood by theoretical physicists – a mental construction that has helped us to make unbelievably precise predictions that have been checked over and over again, via countless different experiments.

Take a place, a region, anywhere in our universe and get rid of everything it contains. And I mean everything.

Strangely, what you are left with is not empty, even though you thoroughly cleared the place of everything it contained.

Does that make sense? Hardly. But nature doesn't care what we humans consider sensible.

Now, please close your eyes.

Why?

Because some things around us cannot bear to be watched, and the vacuum you are about to encounter is one such thing.

---

* Edward Witten is one of the fathers of the so-called *string theory* that you will encounter at the end of Part Seven and, incidentally, the first and only physicist to have been awarded the Fields Medal (the mathematics equivalent of the Nobel Prize).

To make sure you are ready, take a minute to relax and think back to the plane-ride home from your lovely tropical island.

You might remember that you fell asleep not long after take-off. Indeed had you asked him, your weird-looking neighbour would probably have told you that for most of the flight you were snoring your head off.

So what exactly happened during the flight, while you slept for eight hours? What time zones did you cross along the way? And indeed what path does *any* plane take through the sky when no one is actively looking?

All you know about your flight is what you saw before falling asleep and after waking up. You looked out the window and watched as your plane took off from the runway of some faraway island, and you watched it land safely in your home country. In between, no imprint of any flight path was left in your brain whatsoever. You simply don't know what happened.

Now, what if someone told you that your plane took a very unexpected route? Via Jupiter, for instance. Or through Earth, like a neutrino, or back and forth in time? I'm guessing you would have a hard time believing it.

However, whether dreaming or not, you did experience one such weird trajectory in Part Three, travelling 400 years into Earth's future, in eight hours of your own time. So we need to look at what happened more closely.

You now know that for this to happen for real, your plane would have had to fly extraordinarily fast. Indeed it would have had to shoot far into outer space, at near light speed, before heading back towards an Earth that had aged fully 400 years.

In real life, you may find some compelling arguments against such a trajectory, or against any similarly weird paths your plane might have taken, but still: what if I told you that while you slept, your plane not only *did* fly out into space and back again, but that it actually simultaneously took *all the possible and impossible paths* that lead from where and when you fell asleep, to where and when you woke up? Through Earth, and back. Around Jupiter, and back. All of them.

You'd probably never take me seriously again, right?

Good.

That means that you're finally ready to have a look at the vacuum.

Your coffee, the vases, your sofa, your home are all gone.

You are back in a world that only minds can visit, and you are little more than a shadow: completely transparent and yet delineated. Unaffected by, and not affecting, whatever might surround you.

What surrounds you, however, isn't entirely straightforward.

As far as you can tell, there is, well, nothing.

Only darkness, everywhere, stretching to infinity.

By now used to such drastic changes of scenery, you drift gently through what looks very much like a universe emptied of everything it once contained.

The view is rather soothing at first. But soon, admit it, you're bored. With nothing better to do, you start reconsidering what I've just told you about falling asleep on a plane.

Could a plane, could a *real* plane, truly fly in a completely unexpected way? Keeping an open mind about various meandering routes is one thing. But actually flying through the centre of Earth? Or back and forth in time? Come on!

Well, you're right. "Come on!" is the only natural reaction to such a ridiculous thought.

But you should keep an open mind about it all the same, for what may sound crazy for a plane may be very real for a particle.

So let us start thinking instead about a particle, a particle that no one is watching. You picture it having to travel from one place to another, detected by you only at the points of departure and arrival. Now, the same question again: if you do not look, what path does the particle take to get from one place to the other?

Surely it depends . . .

But no, it does not *depend*. For a plane the idea may seem abstract, but for a particle, it is a fact. A particle really does take *all* paths one can possibly imagine, whether they sound reasonable or not, as long as one does not look. Particles move and behave like nothing you've ever seen or experienced in your daily life. You probably got a glimpse of that while touring the inner workings of an atom, seeing that electrons and everything are not just spherical lumps of matter. Now we are approaching an even deeper truth: quantum fields do strange things to particles.

Belonging to a quantum field means that particles really do split into many images of themselves, all the time. And the paths taken by all these images fill in every spot there is in space and time, with you having only a chance, a probability really, of finding a particle at *one* particular time and place, whenever you fancy trying to detect one.

Worse still: before a particle of matter or light is detected, its countless images themselves can split up and become something else before turning back into the particle it was in the first place. Just as light can become an electron and electrons can become light, all the particles in our universe can change into something else when we are not looking. Quantum particles are sneaky little fellows: whatever can happen happens, when nature is left unchecked. And if you do not take my word for it, see for yourself.

•

Something is happening to the endless night of space in which you are floating: a white, doorless cube of a room is beginning to materialize around you, and you soon find yourself inside the room, the walls of which are all covered with extremely tiny, perfectly white detectors. Millions of them.

Right in front of you, in the middle of this doorless room, a vertical metal post as wide as your hand runs from floor to ceiling.

The only other thing in the room is a yellow machine that looks a bit like one of those mechanical devices that throw tennis balls. This queer little robot almost seems to be staring at you through the end of its throwing tube.

Apparently programmed to be polite, it says hello.

It doesn't have a mouth or eyes or ears or anything. Yet it talks, with a rather rusty voice.

"Hi," you reply, just in case, and start to ask a question.

The machine cuts you off, explaining that it is filled with buzzing particles that it will now throw, one by one, to the other side of the room.

If you wonder whether they are particles of light or of matter, the answer is that it might be either – since for what you are about to witness, matter and light would fundamentally behave the same.

Apparently unable to wait, the robot immediately starts a countdown.

"Three . . . two . . . one . . ."

The tube emits a particle and an instant later a bell sounds on the other side of the room. You have the curious sense that the robot is rather pleased with itself.

You lean ever so slightly to one side and see that one of the wall detectors has turned black, behind the metal post.

"First question: how did the particle get there?" asks the robot.

Unfazed by its blank professorial tone, you move to stand in front of the ball-thrower. A straight line links the point at which its delivery tube fired the particle, to the blackened detector. The straight line of that apparent trajectory almost touches the metal post, but not quite.

"That's the path," you announce, raising a finger to point in the only possible direction the particle could have taken.

"Incorrect," the robot responds simply.

"I beg your pardon?" you say, surprised.

"Your answer is incorrect, whatever direction you are pointing at," states the robot, making you reconsider its supposedly programmed politeness.

"But there's only one possible path! I'm looking at it now."

"If you rely on your senses and on your intuition," the machine continues, "then you will continue to answer incorrectly. Every human does so when first entering this room. The rules obeyed by quantum particles are not the same as the ones that rule your everyday life. Your senses and intuition are of no use with particles. Forget about them."

However rude you may feel its attitude to be, the robot is perfectly correct. For despite its rather humble appearance, it fulfils the function in this book of being the most advanced computer in the world – and just as computers are often a scientist's best friend in real life, helping them to visualize their theories, our robot supercomputer will come in useful throughout the rest of this book.

It can simulate anything that obeys the laws of nature as they are known by mankind. The white room you're in, for instance, is a creation of the computer. But everything that happens within the room obeys the known laws of nature.

Now, it may séem that the particle our robot threw flew perfectly straight, but particles belong to the world of the very small, and are therefore beyond the realms of common sense. The computer said you were wrong because what just happened had nothing to do with what your eyes can detect, or how smart you are. The computer is talking about nature, and nature is both intractable and clear on this point: quantum particles do not behave like tennis balls, but like the quantum particles they are. To get from one place to another, they take *all* the possible paths in space and time as long as these paths link their starting point to their end point. The particle the robot shot literally went everywhere. Simultaneously. To the left and to the right of the post. And through it. And outside the room. And into the future and back – until the moment when it hit a detector on the wall.

Now don't worry: you do not necessarily have to understand this. It actually doesn't matter whether you understand it or not, it just *is* the way nature works. Particles that no one looks at *do* travel through all possible paths spacetime can offer. The metal post in the middle of the room changes nothing. In fact it was only there to make the point visually. Take it away and the particle would still travel to the left of where it was *and* to its right.

The detectors on the walls, on the other hand, did make a difference: hitting one made the particle eventually show up *somewhere*.

Next to you, the yellow particle-throwing robot begins to shake and heat up. You wonder whether it might be about to break down, but anticipating your question, it suddenly begins to speak again.

"Everything is in order. I am slowing time down. It takes some energy. The next time you blink, I will throw another particle. You will see what the room would look like if you could witness all the paths a particle travels in order to get from the delivery tube to the wall."

Without thinking about it, you inadvertently blink, and the robot does indeed start another countdown. The flow of time also begins to slow.

"Three . . . Two . . . . . . One . . . ."

The particle leaves the robot in extremely slow motion. At first, it kind of looks like a fuzzy cloud. Moving so as to stand right behind the delivery tube, you then see it split into a seemingly infinite number of ghost images of itself, a wave really, a ripple propagating through the background field it belongs to, travelling in all directions in space and time, including to the right and left of the pole, and through it, and through the rooms' walls, splitting into as many possibilities as your mind can imagine, before suddenly focusing back into one spot on the other side of the room, triggering another detector. A bell sounds, the detector blackens and time recovers its normal rate of flow.

What you've just seen, courtesy of the computer's simulation of a white room, is what scientists believe happens to particles when no one looks at them. When one looks, the whole set of rules changes. When radars keep track of a plane throughout its flight, the plane cannot be at any other position than the one it is detected to be at. Similarly, when one tries to detect a particle, as the detector on the wall did, then the particle isn't everywhere any more, but somewhere. Contrary to a plane with people inside, however, when no one looks, a particle really is everywhere.

•

On the surface, that might sound like the tree that crashes in a forest: with no one to hear it, did it make a noise? And while we're at it, did it really fall?

But we are not talking about philosophy here – we are talking about nature; about how the particles *we* are surrounded by, and made of, behave.

Now why should particles – nature – care if a human is watching them or not? Well, many scientists have pondered just that question. And that led some of them to some crazy answers that we'll meet later on, in Part Six. For now it is enough to say that what you have just witnessed has been shown to be true by countless experiments. Particles are everywhere, and then they aren't any more: in the robot's simulation, the detectors themselves forced the particles it threw to hit the wall of the room somewhere.

"If you are confused, you are right to be," says the robot. "I have shown you that the very act of probing reality changes its nature."

"Say that again?" you ask, frowning.

"Reality changes when you look at it," repeats the robot in a flat tone. "And you are right to be confused about it."

The very small quantum world, it seems, is a mixture of possibilities.

The quantum fields to which all particles belong are the sum of these possibilities and, somehow, *one* possibility is chosen out of all the existing ones just by *seeing* it, just by the very act of *detecting* it, whenever one tries to probe a particle's nature. Nobody knows why or how this happens, but the result is there all the same. Multitude becomes singleness when you interact with the quantum world. Just as, from someone else's point of view, all the thoughts you may or may not have at some point in your life on a given subject suddenly become reduced to one as someone hears you speak it out loud. That's what the detectors at the back of the white room did. They *forced* the particle the robot fired to end up *somewhere* rather than keep being everywhere, dispossessing it of its ubiquitous nature.

As the possible consequences of this begin to dawn on you, you get goosebumps, even though you are still just a shadow. Could this mean that with the right detection equipment, you might be able to make reality your own? Just by trying to detect them, could you make particles – matter itself – move this way rather than that, moulding the whole universe just as you please? Witten said that the universe wasn't made for your convenience, but perhaps he was wrong.

Before you start bragging about it, I'm very sorry to say that Witten was right after all, and that your newfound power is a mirage. You cannot mould the universe, because of all the quantum possibilities the quantum world is made of, it is impossible to predict which one will become real after a glance. This is part of the magic of the fields that make up the universe. The quantum world turns what we thought were certainties into possibilities, or probabilities, for us to probe with experiments, the outcome of which no one can guess with total confidence. Just like flipping a coin or rolling a dice. Scientists have thought that this uncertainty was linked to something missing in their knowledge, but it was proven that it is not thanks to a celebrated theorem published in 1964 by Northern Irish physicist John Stewart Bell. Bell's theorem allowed French physicist Alain Aspect to experimentally show that the existence of possibilities rather than certainties is a property of the very small we'll just have to accept.

All right.

But what does all this have to do with the vacuum you were supposed to probe? Well, that is what you are now about to find out.

The detector-filled white room vanishes, along with the metal post that was in the middle of it and the yellow robot, which didn't even bother to say goodbye.

You are back in the middle of what seems to be the cosmic night, alone, surrounded by nothingness.

You shrink to your mini-you size, and begin to watch the stirrings of something.

It is as if . . . as if a particle (or maybe there were two, you aren't sure) just appeared right in front of you, before vanishing in a puff of light.

There was nothing around, and then there was something, and now there isn't anything any more.

Strange.

And it now happens again. And again. And countless other times, everywhere.

What you are witnessing is the apparently spontaneous creation of particles out of nothing. And before they, for some reason, disappear, these particles travel all the possible paths their quantum freedom allows them to take.

You can accept the last part of that statement. You saw in the white room that that is how unchecked quantum particles behave. But how can they just pop out of nothing?

Well, it is not nothing that surrounds them. There are quantum fields around.

To appear, particles have to borrow some energy from the quantum fields. And since those fields fill in every place in space and time, particles can literally appear anywhere, and anytime. That is the reason why there is no such thing as true emptiness, anywhere in the universe.

You stare further into the darkness and suddenly, as if a filter had been removed from your eyes, the whole truth of it appears to you at once. Particles. Fusing. Everywhere. Filling everything, shooting through a boiling background of fluctuating loops, virtual particles all moving and interacting with each other, appearing and disappearing in puffs of light or energy. An extraordinary firework display occurring everywhere, leaving no spot empty. Pretty much the exact opposite of what you probably once thought was the "nothing" that filled the vast emptiness of outer space.

And this is what scientists call the *vacuum*.

This is what is left when everything is taken out: quantum fields at their lowest possible energy level, with virtual particles spontaneously popping out of them only to move everywhere, before being swallowed back into oblivion.

I will say it once more: there is no such thing as emptiness in this universe of ours.

In a place from which everything has been taken, you might reasonably have expected nothing to be left. But the fact is that, just as you can't take away space and time from any place, you can't take away the vacuum of the quantum fields either.

But if the vacuum is not truly empty – if the vacuum of a quantum field is defined by all the particles that can pop out of it – then a rather legitimate question comes to mind: is a vacuum the same everywhere, or can the nature of it change from one place to another? To use its proper plural: are there many *vacua*?

In 1948, Dutch physicist Hendrik Casimir forecast that for a vacuum defined as above, were all this to be a real fact of our universe rather than just a theoretical fantasy, then not only should there indeed be different vacua around, they would also have a very concrete effect on our world. An effect that could be detected.

Imagine a wall, mounted on multidirectional wheels, separating a room filled with air from another room filled with water. You might expect to see the wall move, gently pushed sideways on its wheels by the water, in the direction of the air-filled room. Now imagine two tiny parallel metal plates facing one another. If left alone, just like the wall separating the water-filled and the air-filled rooms, they should move: they should be repelled – or attracted – towards one another, because of the difference between the vacuum they delimit and the vacuum that lies away from them both.

Why?

For the simple reason that there is more room outside the plates than in between them. Because of this, the virtual particles that pop out of nowhere in between the plates are different from those that appear outside, making the vacua different.

As a result, the plates should move – and it just so happens that they do, as was confirmed experimentally by US physicist Steve Lamoreaux and his colleagues in 1997. This phenomenon is known as the *Casimir effect*.

The Casimir effect confirms that emptiness doesn't exist, and goes further still, showing that different types of vacua occur and can give rise to a force: the force of the vacuum.*

Incidentally, you might notice that you've also just found the solution to a very, very deep puzzle.

As you've known for some time now, all the particles in our universe are but expressions of quantum fields. They are like waves at sea. They are like balls thrown in the air. They are both, particles and waves, born out of and propagating through the quantum field they belong to.

Now, when exploring the very small, do you remember noticing that all the fundamental particles you encountered were always the same? That any two electrons were always exactly identical?[†]

How could that be?

In your daily life, such perfection simply does not exist. Whatever you do, whatever you look at or build or think, there are no two exactly, perfectly identical objects. Or people (even twins). Or birds. Or thoughts. Ever. Even if they *look* similar, they are not identical. So how come all electrons and other fundamental particles are always *absolutely* and *perfectly* identical to any other one of their own kind?

---

* As our electronic devices become smaller and smaller, engineers will increasingly have to take that effect into account.

† And this is also true for quarks and gluons and photons and all the other fundamental particles of all the quantum fields.

The answer is that all the elementary particles, throughout the universe, blossom out of the same background entities that might swallow them back up again at any time: the vacuum of a quantum field. The invisible background seas that fill in our entire universe.

All the electrons are identical expressions of the electromagnetic field, they all pop out of its vacuum and propagate through it. And so do all the photons.

Every time an electron becomes real, it is awoken out of its ghostly lethargy by a kick in the surrounding electromagnetic field vacuum. Every time a gluon appears, it comes from some energy given or taken from the vacuum of the strong interaction field. Every time radioactive decay occurs, the weak field's vacuum is involved and fires off its elementary neutrinos. And the more energetic the vacuum is, the more elementary particles can pop out of it.

All right, we're on a roll, so let's continue: it seems that all the fields behave the same, that they all obey the same rules. Now what about gravity?

Anywhere gravity acts, a gravitational field is actually in effect too, although that field is different, at least for now, because no one knows how it could be a *quantum* field. As you will see later on, no one knows how to make particles pop out of a gravitational-field vacuum without creating catastrophic problems. But if it were possible, then gravity would involve particles that, as with all the other fields, would indeed pop out of the gravitational field to carry its force. These particles are called *gravitons*. They haven't been detected yet, and curves of spacetime are still the best way to explain gravity's action.

But even without them, and even if it perhaps isn't quantum in nature, gravity is nonetheless a field. And that makes the total number of fields used by mankind to describe everything we know so far to be four.

But why four?

Why should there be four fundamental fields?

Why not five or ten or forty-two or 17,092,008 to explain the way nature behaves?

And how about their respective vacua? Are they just cohabitating everywhere without noticing each other's presence? Sounds odd, doesn't it? Wouldn't life be simpler if there were just one field?

It would.

And simplicity is something theoretical physicists are always very keen to find. It even drives their imagination, and that is why they have tried to merge the four known fields above into just one.

One field to rule them all, you might say.

Easier said than done, though.

Each field's elementary particles aren't even the same. And one of them (gravity) doesn't even have any particles that have been detected.

And exciting one field gives different outcomes to exciting another. And they don't involve the same charges. And they don't have the same properties at all, in fact: electromagnetism is long range in its effects and can be either attractive or repulsive, while gravity is only attractive and the strong interaction very short range, and . . .

And yet . . .

To create an alloy out of two different materials, you need to heat them up. Heat them to a high enough temperature and they melt into something entirely new, a new material that unites them both.

For fields to merge, the same idea could work. But an inconceivable amount of energy would be needed – a temperature of around a million billion degrees is required to turn the electromagnetic and the weak nuclear fields into one.

One million billion degrees, that's definitely out of bounds of the nature we know today.

But it may not always have been the case.

In fact, such an enormous amount of energy *did* happen to be available, everywhere, a very long time ago, when the universe was younger, and smaller. And trying to work out, on paper, how nature behaved back then, Salam, Glashow and Weinberg managed to merge the electromagnetic field with the weak field, thus discovering the electroweak field. They found out that under extreme conditions, a single field contained the two fields that today separately rule magnets and radioactivity.

The next step is to unite this new field with the third known quantum field, the strong interaction one, the one that rules how quarks and gluons interact within atomic nuclei. In doing so, we might create something that has been pompously dubbed the Grand Unified Theory. To do that, an even greater energy is needed.

How much more?

A dizzying amount. So much that adding a billion degrees or two doesn't make much difference.

Now, how do we know if all this is real?

How do we know that Salam, Glashow and Weinberg got it right? And apart from the feeling that "one" makes more sense than "three" or "four," how do we know that there really is a Grand Unified Theory out there waiting to be found?

Because in uniting the fields with one another to create a new one, physicists predicted that such a new field should have its own fundamental particles and force carriers. To test this, they have built particle accelerators in which already existing particles are smashed against one another. Within such accelerators, not only are the particles broken apart, showing us what they are made of; the tremendous energy around the collision also excites whatever field lies dormant in this universe of ours.

The maximum energy reached around the impact in such collisions, as of 2015, corresponds to around 100 million billion degrees. That might sound like a lot of energy, but it's worth remembering that we're talking about a *particle* accelerator here. It doesn't accelerate cows or planets, but impossibly tiny particles. In actual terms, the energy produced by these minuscule collisions would barely power the flight of a mosquito. Locally, however, the energy released is enormous. And just as Salam, Glashow and Weinberg predicted, entirely new particles (specifically, the W and Z bosons) were brought into existence – particles that only make sense when considered from an electroweak perspective.

I don't know about you, but such achievements never cease to amaze me.

Now, what about gravity's role in all this? To turn the four fields into one, gravity has to play a role, so why leave it out? Answering that (tricky) question will be the goal of the whole of Part Seven.

But don't be impatient, because with what you've seen so far, you've learnt almost everything there is to know about the matter you are made of, with one big exception: your mass.

Put like that, you may wonder how come you haven't heard about it before: it seems to be quite an important question, doesn't it?

So, where does mass come from?

Stars forge large atomic nuclei in their hearts, as you know, out of small ones.

So do stars also create mass?

No, they don't.

They actually do the opposite.

By expelling the gluons made redundant during the fusion process, the neutrons and protons lose some of their energy, and hence mass, as dictated by Einstein's $E = mc^2$.* This is the origin of the energy that makes stars shine. You've seen this happen. But this also tells you something else: if atomic nuclei lose mass by ejecting their gluons, it means that the gluons *were* that mass. It tells you that part of the mass of the atoms comes from the very existence of the virtual-gluon soups that keep the quarks jailed. In fact, when scientists looked at this carefully, they realized that this "gluon-soup energy" present within all neutrons and protons of our universe accounts not just for some, but for a huge chunk of the mass of the matter we know. A huge chunk. But not the whole of it.

It doesn't tell us, for instance, why quarks and electrons are massive. Or *how* they became massive, rather, for it so happens that they once were massless.

Salam, Glashow and Weinberg showed that a long time ago, as our extremely young universe expanded and cooled down, the electroweak field branched into the electromagnetic and weak fields. But what I didn't tell you before is that for this to have happened, another field had to appear.

Another quantum field, with its own force carriers and all.

These force carriers can't be carrying any of the forces you've already met, and there isn't any other force to account for . . . so what do they do?

Well, they gave a mass to some particles and left the others massless. The photons and the gluons, for instance, did not feel its

* Remember: the more protons and neutrons there are in an atom's nucleus, the fewer confining gluons are required by the quarks to be kept within their jail's boundary.

presence and they still do not feel it. They can travel through its field without noticing it. So they stayed massless and they still travel at the speed of light today.

But quarks and electrons and neutrinos did notice its presence, and became massive. Because of that, they can't reach the speed of light any more.*

Again, how do we know this to be true? How do we know a mysterious field is responsible for the masses of these particles?

Well, like all fields, this new field should have its own fundamental particles.

As expected, however, they are not easy to see or detect.

According to calculations, for this field to be awoken and give birth to its fundamental particles, a tremendous amount of energy is needed – even more than for the electroweak field itself. Still, in 2012, as amazing as it sounds, scientists managed to do exactly that at the LHC, the most powerful particle accelerator of the European Centre for Nuclear Research, near Geneva, in Switzerland.† They detected a fundamental particle that belongs to this field. It was the missing piece of the puzzle: the origins of all the known mass of our universe, be it due to gluons or not, were then known.

It indeed confirmed that physicists had been on the right path all along.

The media called this detected particle the *Higgs particle* (although there may be many different sorts of Higgs particles), and the field it was extracted from is known as the *Higgs field* or the

---

* Neutrinos indeed do have a mass, but their mass is so minute that it eluded everyone until the extraordinary ingeniosity of Japanese physicist Takaaki Kajita and Canadian physicist Arthur B. McDonald proved it to be non-zero. They jointly received the 2015 Nobel prize in physics for it.

† LHC stands for Large Hadron Collider. All the particles that feel the strong interaction field are called *hadrons*. Protons are hadrons and the LHC is basically making protons collide very energetically.

Higgs–Englert–Brout field. British theoretical physicist Peter Higgs and Belgian theoretical physicist François Englert jointly received the 2013 Nobel Prize for this discovery (which they'd predicted more than forty years earlier, with Brout, who sadly passed away in 2011*). In short, they had discovered how some of the mass came into being, 13.8 billion years ago, as our universe cooled down. A very impressive feat for them, and for mankind.

Because this discovery made the headlines, it may nonetheless be noteworthy to underline again that the Higgs field is not responsible for the mass of *everything* that we are made of. Just some. Most of the mass of the neutrons and protons comes, as we said above, from the force that confines the quarks within their boundaries, from the quark-gluon soup that lies in there. If the Higgs field was suddenly turned off, then quarks would become massless and we'd die. But the mass of the proton and neutron would barely change.

Now that the strong field's role in our being massive is vindicated, now that you know where all the mass of all the matter we know of comes from, think back to all those particles that you saw popping out of the vacuum, earlier on in this chapter. You saw them . . . but you shouldn't have. Nature doesn't let particles appear on their own like that without paying a price.

This price, as you are now about to see, is the existence of a new type of matter named *antimatter.*

---

* The Nobel Prize rewards only living scientists.

3 | *Antimatter*

For almost all of Earth's history, its surface was mostly unknown to human beings. Today we have easy access to satellite images of our entire planet but, as recently as a few centuries ago, when only a few patches of European, American and Asian soil had been charted by the people who lived there, no comprehensive image of the world prevailed. Intrepid explorers from many civilizations therefore had to leave the safety of their shores and sail through gales and storms to figure out what, if anything, lay beyond their homelands. One after the other, they found faraway landmasses upon which no fellow man or woman had ever stepped. They found other civilizations too. Small pieces of rock surrounded by water became called islands. Large ones, continents. Each such discovery increased the realm of mankind and, at the same time, tended towards making our ancestors grasp a very simple fact: we all live on the surface of an incredibly rich but rather small ball that drifts through an immense universe.

Decades went by.

Through a mix of violence, greed and curiosity, Earth became better known; the unknown gradually changed from being somewhere beyond the horizon to everywhere above our heads. Space became the new mystery everyone could wonder about, just by looking up. But the distances out there are mind-bending. As this book is being written, human-made satellites have been sent several hundred million miles away from Earth to try to figure out the origins of the water, and maybe the building blocks of life itself, on our planet. Exploration is not just about sending humans on perilous adventures any more. Robots do that for us. But as excitement about interplanetary travel is on the rise again, is it possible, in the early twenty-first century, to stay on Earth and still be an explorer?

Of course it is.

One could aim for the bottom of the oceans, an environment so hostile to our technology (not to mention our bodies) that fewer people have made the plunge down there than have set foot on the Moon.

Or one could have a different approach altogether, and do science.

Although science might not be as glamorous as sailing a caravel or piloting a rocket ship, it can carry you *anywhere*. From the bottom of the seas to the edge of our visible universe. And beyond. As you've probably noticed while reading this book, your mind can take you to places forbidden to your body, and to places where no one has ever been before. While delving deep into the nature of space and time, or the quantum behaviour of particles and light, no two readers of this book will have made exactly the same journey – none will have pictured the exact same things. In creating galaxies and virtual particles of light in your mind, you entered the world of theoretical research, a world without limit.

No one ever knows in advance in which direction an undiscovered island, or a continent, may be found. And many an explorer must fail, in order to pave the way for a great discovery. Luck indeed happens, but it isn't reliable. Building upon past discoveries, however, is. The same is true of science, and the discovery of antimatter follows this ancient, pioneering path. One genius of a man opened everyone else's eyes to the following amazing fact: the matter we are made of, the matter that makes up the planets, the stars and the galaxies themselves, *is but half the matter there is* – and he did not figure this out by sheer luck. He was building upon what had been done before him. Explicitly: on Einstein's work on how things move when they move very quickly, and on the curious ways quantum particles behave. That man was Paul Dirac. He created the idea of a quantum field, and consequently discovered antimatter. Dirac was a British scientist who, between 1932 and 1969, held the Lucasian Chair of Mathematics at Cambridge University, one of the most prestigious scientific chairs in the world. Isaac Newton sat on it between 1669 and 1702, and so did Stephen Hawking, between 1979 and 2009.

So what is this antimatter?

You already know what $E = mc^2$ means: that mass can be converted into energy, and energy into mass. With quite a high exchange rate. And as you have just seen in the previous chapter, that energy can be borrowed for a short amount of time from the vacuum, from the fields, to create particles.

Now back to your mini-you.

•

You are still in an emptied universe, surrounded by a vacuum – specifically a vacuum of the electromagnetic field.

Right in front of you, an electron pops out of it.

Why? Because it can. So you see an electron appear. *Pop.* Just like that.

A moment ago, there was nothing but the vacuum. Now there is an electron, and the electron has a mass. The very fact of it having appeared means that some dormant energy has been transformed into that mass. That's $E = mc^2$ in action. That's easy.

But the electron also has an electric charge. Which rather begs the question: where does that electric charge come from?

Mass comes from energy, and mass and energy are equivalent, so the appearance of mass out of borrowed energy is a balancing process. It's just a change of energy from one form to another. But the electric charge is a different problem altogether. After the electron appears, a negative electric charge also appears. Before, there was none. After, there is one. And that, surely, is not acceptable. As I mentioned at the end of the previous chapter, you can't create something out of nothing without having to pay a price for it. That never happens in real life – I can hear you sigh – and for once it is the same in the quantum world.

So, what do we do with this charge? Do we just turn a blind eye to it?

We can't do that, because there is far too much of it. Every electron in the universe carries a charge, and many other fundamental particles do too.

So where does the charge come from?

Well, the easiest answer often being the right one, here it is: an electron never appears alone. It must appear alongside a particle that is identical to it, except for its charge, which is opposite. Such a particle is called an *anti-electron*.

It was introduced so that the charges of all the electron–anti-electron pairs that have ever been created adds up to zero. No need to invoke $E = mc^2$ any more or indeed anything else. Such a phenomenon doesn't violate any laws: the total charge was zero before the electron and anti-electron appeared, it still is zero afterwards.

That is what Paul Dirac, somehow, brilliantly figured out.

What's the big deal? you'd be forgiven for wondering.

Well, a particle that is exactly similar to an electron, with opposite charge, was not known to exist at the time. No one had ever seen an anti-electron.

Today, we detect them everywhere.

The process by which an electron and its anti-self appear out of nothing is called a particle–antiparticle *pair creation*, and the opposite process also exists: when an electron meets an anti-electron, they *annihilate*, they disappear, *puff!* gone, their mass transformed back into energy, into light, in an instant.

Electrons and their anti-selves are created out of the electromagnetic field, and they melt back into it when they annihilate.

Now, since electrons exist on their own, and since they were all created out of the electromagnetic field during an electron–anti-electron pair creation, it follows that anti-electrons should also exist on their own. And indeed they do. But they are not to be found everywhere.

In 1928, Dirac had called the anti-electron a "hole in the sea," the sea being what we now call the electromagnetic quantum field, because it corresponded to some charge that was missing.

His "hole," the anti-electron, was discovered experimentally five years later in 1933 and Dirac received the Nobel Prize in Physics that year, for his extraordinary insight. His theory of fields encompasses all those very fields you've been seeing everywhere since you started to explore the world of the very small, and discovered anti-matter.

It was US physicist Carl D. Anderson who actually *detected* Dirac's anti-electrons for the very first time. Instead of calling them anti-electrons, however, Anderson gave them a new name: *positrons*, a name that is still in use today. Anderson got the Nobel Prize for his detective work three years later, in 1936.

With this, antimatter was born.

I said earlier that half of all matter was antimatter. But if it's only anti-electrons, then it's not half of *everything*. What about the anti-quarks and anti-light and anti-gluons?

Well, what is true for the electron is true for all particles.

They all have their anti-selves.

Anti-quarks exist, and so do anti-neutrinos and anti-photons. But some particles, ones that do not carry a charge, can play both sides and be their *own* antiparticles. Light is a good example: because photons and anti-photons carry no charge, they are identical.

So why don't we see all the other antiparticles around us, everywhere we look?

The answer is that they *are* there, around us, around you, but not in large quantities, because whenever one appears, it only lives for a very, very short time. Remember, any antiparticle that bumps into its particle counterpart immediately annihilates with it, only to disappear into a puff of energy and light, according to $E = mc^2$.

Somewhere else in the universe, however, a whole world could be built out of antimatter. An anti-world, if you like. No one knows if such anti-worlds exist, but if they do, and if you happen to end up facing someone like you in outer space someday, don't shake hands. You and your anti-you would become a bomb and immediately explode. Violently.*

Still, there is some antimatter around nonetheless. Even inside you, right now.

Every time an instance of radioactive decay occurs, some antimatter is created, only to annihilate with its matter counterpart to become a ray of light so powerful that it usually shoots through your body, without you, or anyone, noticing it.

Your eyes can't see these rays because, as we discussed earlier, your eyes simply never needed to evolve the capacity to detect them. But what your eyes cannot see, technology can – and some quick-witted engineers have managed to turn that discovery into efficient medical diagnosis-and-research tools. PETs are an example. They are used in hospitals. PET stands for positron emission tomography. Doctors inject liquid "tracers" into your body that are themselves radioactive, and which emit a positron when they decay. The positrons then annihilate with electrons in their path, turning into powerful gamma rays that are detected outside your

---

* How violently? Well, according to $E = mc^2$, to release about three timesas much energy as the nuclear bomb dropped on Hiroshima, one only needs a *single gram* of antimatter annihilating with its matter counterpart. A 150-pound you–anti-you encounter would therefore be equivalent to 210,000 such nuclear bombs. Quite a handshake you have there.

body by the PET machine, to reconstruct a 3D image of how your body works. It's quite brilliant.

All right.

You now know about fields and their vacua.

You know about their possible unification.

You know about mass and charges and antimatter.

And this means you are ready to travel beyond what you've seen in Part One, to the Big Bang, and beyond still, to the origins of space and time.

So I'd take a deep breath, if I were you, before turning the page.

For years, without thinking about it, you probably took for granted, quite unconsciously, the fact that this universe of ours is mostly emptiness, and entirely fixed and unchanging. Unlike our ancestors you might have heard about the Big Bang, but you perhaps never really thought about what that term might actually *mean*.

In fact, in many ways we are all like those fish swimming in the sea. Except, as you now know, that we are not swimming in a sea made of water, but in the many seas that our friend Dirac charted, seas that are called fields, and that fill the entire universe; fields of which we are a mixed and rather complex expression.

Thinking about it, you reckon that it kind of makes sense, that everything becomes much easier to understand this way: time, mass, speed, distance, all intertwined within these fields.

The universe is huge. Unbelievably large volumes stretch between any two stars, galaxies, clusters of galaxies. But there are no voids. There are only fields that allow for distant objects to have an interaction, by exchanging particles, their so-called force carriers, without ever touching each other.

Fields link everything to everything.

There is something almost reassuring in that thought.

And as you are about to rewind the whole history of our universe up to the birth of space and time, you may wonder: over the course of human history, could all those shamans and holy men and tripped-out people who have shouted, cried, sung, written, painted, danced, over the ages, that "One is All and All is One" have been right?

Well, in a far-fetched sense, maybe.

But they certainly did not know *why*.

Our supercomputer of a robot does though, and it has reappeared.

Once more the bright-yellow tennis-ball throwing machine is right in front of you. It still doesn't have a face and looks at you rather blankly with its particle-delivery tube, but you now know better than to see it as "just" a mechanical machine.

Feeling strong and buoyed with the confidence of all the knowledge you've gathered so far, you prepare your mind to stretch itself once again, to picture the whole history of our universe.

A metallic voice chimes out of the void:

"Are you ready?" it asks.

You know it is going to bring you to the origin of space and time but the robot doesn't leave you the time to answer and an instant later, you are with him in the sky. Above a house. Your house.

The computer has brought you back from wherever you were to above your hometown.

And you are now both shooting upwards.

You cross the different layers of our planet's atmosphere and reach space again, where you settle above your home world, to face outer space.

"I am going to have you fly through the best simulation ever made," announces the robot. "When programmed, as I am, to obey the laws of nature that have been unravelled so far, even the most powerful supercomputers on Earth struggle to achieve what you are about to see."

"Then let's go!" you exclaim, feeling the excitement of the journey build, itching to go beyond what can be seen and cross all these interleaved layers of pasts that pile up around the Earth.

You know that if you wanted to reach a star normally, with your body rather than your mind, then you would have to spend some time travelling, and the star would not be as it is *now* when you reach it. It would have evolved. Just as if you wanted to travel to Paris now, it would still take a few hours to get there. The Paris you'd reach would be different from the Paris that was there when you set off. People and cars and clouds and raindrops, nothing would be in the same place any more.

While travelling to a faraway star in a faraway galaxy, the difference would be even bigger. By the time you reached your destination, the universe would have expanded. The cosmic microwave background, the overall temperature of our universe, would be colder, the surface of last scattering further still. Travelling normally, however fast, you'd never reach the past.

So how can the computer simulation propel you into the past – and the very distant past at that?

The answer quickly dawns upon you: to find yourself in the universe as its infancy unfolded, to see it happening, you shouldn't move. You should just let time run backwards, and that is exactly what starts to happen.

Without moving, you begin a new journey, travelling backwards in time through the history of our universe to reach the Big Bang and beyond, from the point of view of where you are.

Showing a sensibility you had not expected of him, the computer's robot avatar even kindly fades away so that his presence doesn't hinder your sight.

In a blink, you are 7 million years in the past.

The surface of last scattering, the surface delimiting the end of the visible universe as seen from Earth, is already slightly closer – and it is filled with a slightly hotter cosmic microwave background radiation. But 7 million years isn't much compared to the 13.8 billion-year history of our universe, and nothing out there in the sky is particularly different from a moment ago. The Earth beneath you, however, is. Down there, there are no towns or cities, no twinkling street lights. The first humans are just beginning to differ from the great apes. Your distant ancestors are quite hairy, hunting beasts. Humankind has indeed come quite a long way . . .

Another blink and you are 65 million years in the past.

The dinosaurs have just been wiped out by a mixture of violent volcanic eruptions and a cataclysmic collision with a 6-mile-wide asteroid, leaving only small mammals alive, some of which will one day, after many successive evolutions, become the hairy ancestors you just saw, and then us.

Another blink and you are more than 4 billion years ago.

The Earth just got hit by the Mars-sized planet that took a chunk off it, to create the Moon. The microwave background radiation is definitely starting to be hotter, and the surface of last scattering now really does look closer than before. The whole visible universe, as seen from when you are, is less than 70 per cent of what it will be in 2017.

You rewind another couple of billion years.

The visible universe is less than half the size you started from. Earth does not yet exist. In its place, stars are dying right in front of you, extraordinary explosions that spread the matter they were made of through outer space. In a few hundred million years' time, this dust and debris will gather into huge clouds, and gravity will drive the formation of at least one new star, the Sun, and its planets.

Another blink, and you are 5 billion years before Earth was born, 9.5 billion years before *you* were born.

Your visible universe is less than 25 per cent the size it will be in 2017. The surface of last scattering is much closer to you. In between you and this wall, galaxies are forming around some gigantic black holes, sometimes meeting in collisions of unimaginable magnitude.

Another blink and you are 13.7 billion years ago.

You are still where Earth will someday be, but the visible universe, the universe that surrounds you, is now less than 0.5 per cent the size of the one you started from. You are within the Dark Ages of our universe.

The Dark Ages you travelled through in Part One of this book were cold because, back then, you flew through what they look like as seen from Earth in 2017, after more than 13.7 billion years of expansion.

13.7 billion years ago, however, things were neither cold, nor dark. And you are there now.

The first stars haven't ignited yet, so none of the matter you can see has been processed through nuclear fusion in star cores. You are therefore surrounded by the smallest atoms that can be: hydrogen, mostly, and helium. And the radiation that glows throughout – the cosmic microwave background radiation – is not microwave radiation either. You can *see* it with your eyes. It is the light that originally filled our universe, a light that shines brightly everywhere, a light that will only become microwave radiation much later, after several billion years of our universe expanding.

Another blink and you are 100 million years earlier, 13.8 billion years ago. The surface of last scattering, the surface at the end of the visible universe, is now one light minute away from you, meaning that your visible universe is just one light minute deep, less than an eighth of the distance separating Earth and the Sun.

The whole universe has been transparent for just sixty seconds. And it is hot.

5,400°F, everywhere.

Still in the Dark Ages, but everything around is so luminous that you wonder if the description really fits.

You pause there.

In a moment, the computer will start rewinding time some more, albeit at a slower pace, and you will enter a strange, literally invisible place. One further minute into the past and you'll have started what sounds like the ultimate journey . . .

The surface of last scattering is right in front of you.

You take a deep breath, ready to cross it, to travel beyond the wall, to reach the unseeable.

Time rewinds . . .

And you're through.

You've entered a part of our universe's past that will never be seen with light.

Indeed, you can't see anything any more.

Light does not propagate here. There's simply too much energy around.

But you know what to do.

You immediately switch into yogi mode and, to your great surprise, you realize that the universe beyond the surface you've just crossed is *big*.

And old.

At least 380,000 years old.

Your journey is far from over.

You concentrate once more on what is around you, on what is happening now, behind the wall at the end of the visible universe.

The surrounding temperature is 9,000°F. All the electrons that will one day bind to loose atomic nuclei to become hydrogen and helium are here on their own. Photons bump into them, exciting them, before being emitted back, only to bump once more into another electron. The electromagnetic field is so filled with energy that all its fundamental particles turn into one another in almost no time.

Another blink and you've run tens of thousands of years backwards from the moment the universe became transparent.

You are surrounded by a dense broth of particles, a mixture of all the excitations of the quantum fields, their elementary particles and their force carriers. All bumping into each other, none of them managing to travel at all. There is too much energy around. They appear. They collide. They disappear. And as time keeps rewinding, as the universe continues to shrink, as the energy density rises, everything gets more and more violent.

Still, you try not to get confused, and focus on your journey backwards in time. You are pure mind, and in yogi mode, travelling through what looks like a very, very realistic simulation. The universe keeps shrinking, its fabric, spacetime, bent to prodigious levels. Gravitational waves are everywhere. Nothing you know or imagine would stand such crushing and shearing power.

For a split second, you wonder why you have not heard more about gravity at this stage, but you have no time to think about that. You have gone back in time some tens of thousands of years more and you are now surrounded by an unimaginable inferno. Your virtual heart starts thumping harder and harder as the temperature and pressure and the effects of gravity on what you see rise to unbelievable levels.

You now are 380,000 years before the universe became transparent. Looking through a telescope now on Earth, seeing back over 13.8 billion years, you are 380,000 years beyond the wall that marks the limit of the visible universe.

Looking at it the other way round, you are about three minutes away from what one might call the birth of space and time.

As time continues to run backwards, even the atomic cores are breaking down, leaving all the neutron and proton quark jails free to move on their own. The strong nuclear force itself is overwhelmed by the ambient energy. Protons and neutrons, those sturdiest of constructions, even enter a frenetic dance in which, pounded by quark-made force carriers, the protons, bewildered, turn into neutrons, disappearing from the universe.

The temperature?

100 billion degrees.

Everywhere.

But you do not stop.

You keep moving. Rewound second after rewound second, all the particles of light that surround you now turn into matter and anti-matter pairs. Everywhere. And there seems to be as much of the latter as of the former. How did one end up becoming predominant, then? you wonder, in a semi-trance. Something special must have happened around here, for that equilibrium to have been broken. A mystery that may even be solved this year, or next, as the updated and boosted particle accelerator LHC (switched back on at CERN in June 2015) starts revealing its new finds.

You wish you could stay here just a little longer to figure it out for yourself and scoop CERN, but you are not in charge here and you are now cruising through a universe filled with a soup of such formidable energy that everything is shaken to the extreme, gravity bending and crushing, fields excited to levels beyond sanity.

It is not the weight of a star that gravity, through the bending of space and time, imposes upon every field around here, but the energy of the whole universe squeezed within a sphere 100 light years in diameter.* Such a sphere, centred on today's Earth, would contain no more than 5,000 stars. Back then, it contained the

---

* If you wonder (and rightly so) why the universe is 100 light years wide rather than a couple of light minutes, you will have the answer in Part Five.

energy to build hundreds of billions of galaxies that each contain hundreds of billions of stars. Not to mention the dust.

Still, as much as you would like to look at it all, you keep flying against the stream of time.

You now are about a millionth of a second away from your final destination.

The temperature has reached 100 million billion degrees.

With so much energy around, even the quark-jail guards, the gluons themselves, can't keep their prisoners confined. The neutrons break down. The quarks, now free, start to interact with their anti-selves, turning into pure energy.

As you look around, you realize that the difference between matter, light and energy is now completely superfluous.

Fields that had been separate entities all the way from Earth's time to here, fields that on Earth described everything you could think of, through different forces, are now merging into one another, as expected. The electroweak field is active. As some of the old particles you are used to seeing everywhere disappear, new ones, fundamental entities belonging to the electroweak field, pop out everywhere. The Higgs field vanishes. And with it, the massive Higgs particles that remained hidden to human knowledge for so long.

The particles you now see are the W and Z bosons we met earlier, the force carriers of the electroweak field.

There is so much energy around that these particles, so hard to create on Earth, are everywhere.

The universe is now 100 billion billion degrees hot and the laws of nature begin to differ noticeably from those you've experienced throughout your life.

Quarks and anti-quarks disappear.

Gluons are swallowed within the background field.

A thousandth of a billionth of a billionth of a billionth of a second after what is nothing less than the presumed origin of space and time, an event we might call the *Beginning*, what will one day become our whole visible universe, is now a sphere 33 feet across, and it keeps shrinking.

Everything it contains is now heated to a fantastic billion billion billion degrees. And as it keeps heating up, all the fields that make up all the matter we are made of unite to become the grand unified field.

Only gravity lies outside this unification of the forces.

Being so close to the Beginning, you start to believe that not much could still happen.

In fact, you've just reached what has been called the Big Bang: the moment the energy stored in the grand unified field began to turn into particles.

But astonishingly, even though experimental physics has never reached such a place, the computer seems unwilling to stop, if only to show you that the history of the universe does not begin there. Indeed, to your great surprise, as time keeps rewinding, the whole universe's matter and energy suddenly vanish – and contrary to what you might have expected, everything cools down tremendously while all the available energy is turned into yet another field, a field you have not met before, a field filled with its own particles.

It is called the *inflaton field*.

It is considered to be responsible for our universe's initial expansion.

As crazy as it may sound, everything now suddenly speeds up again and the whole universe mind-bogglingly collapses in on itself at an unimaginable rate, dragging you in.

In less time than it would take light to cross the core of an atom in your kitchen, the whole universe shrinks from about 33 feet in diameter to a size billions of times smaller than a proton.

Scientists have called that period the *cosmological inflation.*[*] You've just been through it – backwards. Beyond, there is no more matter, no more anything.

All the known fields are gone.

The laws of nature bear no resemblance to the ones you've experienced throughout your life, or throughout your journey up to this point.

It is somewhere around here that the three forces or fields that will go on to rule all the known matter and antimatter of the universe in the present day, including the matter you are made of, were once, it is believed, merged with gravity.

You would like to keep going, to keep rewinding further beyond the Big Bang, all the way through the birth of our universe, but there is something wrong.

The notions of space and time you've used until now do not apply any more.

The gravitational bending of spacetime is too strong. Quantum effects are too strong.

With no time, with no space, with no spacetime, you cannot travel any more. Travel does not actually make any sense in such circumstances.

You haven't reached the Beginning, and you do not even know how to think of a way to get there.

How frustrating.

And you suddenly wish you could look at all this from outside, for you've always stayed *inside* the universe so far. But the very concept of outside doesn't seem to make any sense either.

---

[*]   You'll hear more about inflation in Part Seven.

What you've reached here is the surface of another wall, a wall of a different nature from the surface of last scattering that limits what you could see from Earth. This is a wall impenetrable not to light, but to modern knowledge.

Beyond lies the realm of *quantum gravity*, where all the known fields of nature may be united into one, in a quantum way.

There, our universe becomes a mystery in which twenty-first-century sciences, beliefs and philosophy are intertwined. In a way, this is where our knowledge stops, and where pure theoretical research takes over. To travel beyond the surface of last scattering, you cannot use light-gathering telescopes, but scientists have built particle accelerators that allowed them to reach the temperatures and pressures they expected to find beyond, and it worked. Scientists figured out new laws, and they managed to crawl back along the flow of time, albeit in an indirect way. Gravitational wave detectors, however, now allow us to detect ripples propagating through spacetime itself. These waves care about no wall. So the tantalizing prospect of one day detecting signals from the distant past you've just travelled across, where plenty of such primordial waves must have been emitted, is not a pure fantasy anymore. But to travel beyond the wall of quantum gravity, also known as the *Planck era*, is a different proposition altogether. No one is even certain how to *think* about what lies beyond. Our whole visible universe was so tiny back then that to probe it in your mind, you need a theory of the immensely big turned minuscule, a theory in which the quantum laws – with their quantum jumps and all – are applied to the universe itself. You need gravity *and* quantum effects. You need quantum gravity and more. And we don't have it. We do not have that working framework. So you cannot go further. In fact, you are not even allowed to infer what lies beyond that Planck wall, neither in space nor in time, because these two notions do not make any sense there. When scientists say that our universe is 13.8 billion years old, they mean that 13.8 billions of years have passed since the space and time you are used to started to make sense, since *space-time* made sense. And that moment occurred something like 380,000 years beyond the surface of last scattering, 380,000 years before the cosmic microwave background radiation filled outer space. And that moment occurred some millionth of a billionth of a billionth of a billionth of a second before the Big Bang. In the end, scientists can rightly say that such a time has elapsed since the origin of space and time. But it does not mean our universe started there. Nor that it is the only universe there is. Nor that it is the only one to have ever been.

•

You are back in your living room, back on your battered old sofa, and you are struck by a sensation so profound that you grip your sofa's armrest.

You've travelled through space and time. You've seen galaxies. You've seen stars. You've seen fields. You've seen how gravity works, how its effect on the shape and fate of spacetime depends on what the universe contains.

Yes. You did all that.

And now something amazing is starting to happen to you, as if you were on the verge of making a groundbreaking discovery . . .

Thoughts are racing through your mind. You feel like a kid again, a kid who suddenly realizes that the world can be understood, that the world, somehow, up to some point, *has been* understood, and the computer showed it all to you . . .

You learnt from Einstein's general theory of relativity that you could figure out the entire history of the universe, if only you knew what it contained.

You now know that the contents of the universe consist of quantum fields that move and evolve and interact, fields that today are three but were once *one*, a long, long time ago.

Those fields are the mothers and fathers of all particles and anti-particles of our universe, and they are the reason why all elementary particles are exactly the same, whether here, inside your body, or in any other galaxy, in the present or in the past.

And all this can only mean one thing.

It can only mean that, potentially, you've become a god.

Yes.

A god.

You know about gravity.

You know what lies inside the universe.

Put the two together, and you know everything.

The universe's history.

Its past.

Its present.

Its future.

You are a god, almost by definition.

Your whole face lighting up, you immediately pick up your cell phone and dial the number of the only person you can think of right now.

"Who is this?"

The voice on the other end of the line sounds suspicious. It's your great-auntie.

"It's me!"

"Oh! Hello, dear. How are you doing? Are you feeling better?"

"Better? I feel amazing!" you exclaim.

"That's nice to hear. Has something happened?"

"I've been travelling, and learning about the universe, and . . . well I know it sounds silly, but I can create and evolve a universe like ours, just by using my imagination. This must be what it feels like to be a god."

Your great-auntie pauses.

"I see," she says.

"What do you see?" you ask, wondering why she so obviously doesn't share your enthusiasm.

"Nothing. Nothing. It's just that, well, I've heard that one before."

"You have?"

"People like to play at being gods, don't they? Do you remember my dear friends Kati and Gabi?"

"I don't, no, but listen, what I'm—"

"Let me finish my story, dear. So, Kati and Gabi and I went to the archery field last weekend. Kati and Gabi are quite fond of shooting arrows, you see, and they taught me this: with some quite rudimentary knowledge of how our world works, it seems that knowing how an arrow is shot and where from, you can tell where it will fall. Fascinating, isn't it?"

"Sure, Auntie, that's ballistics, that's Newton's law."

"Is it? Well, that's good to know. And can you apply that to the whole universe?"

"What?"

"Do you have something to start from, or with? Do you have something to which you could apply your ballistics or whatever laws of nature you've apparently found?"

"I . . . Some sort of initial condition, you mean?"

"I don't know. Do you want me to ask Kati and Gabi to give you a call so you can talk about it with them? They're really rather good at this sort of thing."

"No no no! No need for that . . ."

"Okay then. Will you call me back when you've found your initial condition?"

"I . . . I will, okay."

"Thank you for calling me, dearie. You are a sweetie. Bye now."

With this, she hangs up.

And as you stare blankly at your phone, let me tell you this: as you've probably figured out yourself, she's right. To understand something about the universe, you need two pieces of data. The first one is a law, or a set of laws. The second one is an initial condition.

To apply to the universe as a whole the same ideas you've had so far, to know its whole destiny from scratch, then all the laws in the world wouldn't be enough.

You'd still need a rock-solid initial condition, a state to which you could apply the laws of evolution. And you just don't have it. And to make it worse, how can you even be sure that the laws of gravity and of quantum fields as you know them even applied at the very beginning of our universe?

With a sigh, you sit back in your sofa and grab your mug of coffee with the feeling that some important data is still missing somewhere . . .

Space.

Time.

Spacetime.

What remains to figure out about them, that you have not already seen?

Particles. Force carriers.

Fields.

Gravity and its waves.

Haven't you experienced all there is to know?

And why are you feeling so fazed?

You open your eyes.

To your great surprise, you are not at home any more, but instead sitting in a cramped seat on a strangely familiar plane.

Seat 13A, to be precise.

The other passengers are lining up in the corridor, getting ready to disembark.

You stare out of the little window, confused, but there is no doubt about it: you really are back inside your time-travelling plane. And it has just landed, in 2416. Finding it hard to think straight, you stand up, follow the other passengers out, and find yourself walking through long, seemingly endless glass-walled corridors overlooking the sea.

Why are you back here?

A moment ago, you were at home. You had just called your great-auntie after having travelled throughout the known universe.

Centred on the earth, you remember, there is a sphere 13.8 billion light years in radius which contains all the pasts mankind will ever be able to collect using light. Beyond, another layer of reality existed for 380,000 years. And beyond that? No one knows.

As you keep walking through yet more corridors, a bright 2416 Sun shining its 8.3-minute-old rays upon the future Earth, an unfathomably deep feeling of loneliness suddenly hits you.

What is the point of all this?

How can this universe of ours be so big, and we so small within it? Are we for ever doomed to be lost in space and time, and tortured by our own awareness of that fact? Or are we humans at the beginning of a long technological journey that will some day bring distant worlds closer? Is that what is happening here? Are you about to see one of the many futures our planet might reach, a future in which far and near are no different, a future in which pasts and futures are just travel directions for our descendants to choose from?

Time travel has been a human fantasy for ages, but you've never heard of anyone actually making the trip.

Stephen Hawking once held a party for time travellers, at midday on June 28, 2009. To make sure only time travellers showed up, he didn't send the invitations until after the party. No one came.

So what is this new journey of yours supposed to tell you, *you*, an insignificant organism lost in the immensity of space and time?

The glass corridor you've been walking through eventually curves into the lobby of a huge airport, or maybe we should call it a timeport. Hundreds of people are queuing to get through what seems to be some form of customs. The hall is very bright. Light pours in from gigantic window panels looking out at skyscrapers rising from the sea. As you join one of the many lines, blending with the other passengers, you suddenly fear that what you are experiencing now is not a dream, but real, that what you dreamt was being back home. Understandably this makes you anxious.

If this is real, then what happened to your past?

If you've really travelled 400 years since take-off, is the past you left behind still somewhere out there? Could you, if you wished, travel back to live that past life through, or is it gone for ever? And the dear friends who sent you home from the island, are they all long dead? It dawns on you that this must be the case, that you squeezed past their time towards the now.

The interplay of time and space may be tricky to grasp, but you find it hard to imagine that several lives could be lived by the same person simultaneously, in the same universe, while he or she was aware of them all – even though fields seem to allow exactly that in the case of particles which no one is looking at.

What is possible for single particles appears not to be possible for collections of billions and billions of them, such as the human body. As your mind reflects on that fact with a certain sadness, you become almost physically aware of the unbridgeable gap that now separates you from all the people you loved, and sorrow grips your heart.

But there is a certain solace in what you have seen so far. Your loved-ones' past lives have become a succession of images moving through space and time. All the light and other massless particles that once bounced off their bodies, or interacted with them in even the smallest way, created a memory of their existence, an image, a shell that spreads at the speed of light from the Earth towards far-away unknowns, small ripples on invisible but ubiquitous fields. And since you've travelled 400 years into the future, the visible memory of their lives presently washes over planets and stars that lie 400 light years away from Earth, and their image will keep moving away, spreading further, maybe falling here and there in some light-gathering device aliens might be using, for as long as our universe will be.

And what about the matter they were made of? What about those atoms that were born billions of years ago in the core of long-gone stars, before coming together to form the bodies of your friends and loved ones? All their trillions of trillions of particles are now scattered throughout the world. . . You may even be near one right now. All particles are but one anyway.

Perhaps we are not that small in the scheme of things after all, you think on reflection. Our image is here and will for ever stay, and there is comfort in knowing that the memory of our lives will always be there, travelling among the stars.

Time, space and fields make us all belong to an extraordinarily large reality.

Spreading your arms to feel the fields you are made of, lifting your hands high in the air to see them climb the invisible slope the Earth creates in its surrounding spacetime, you begin to understand how interconnected all pasts and presents and futures may in fact be.

"Is everything all right, sir?" a uniformed woman suddenly asks.

Snapping out of your reverie, rather embarrassed not to have seen the woman approach, you barely manage to mumble that yes, you are fine – but some things in life never change. Even in the year 2416, everyone immediately feels like they're guilty of something when faced by a trained customs officer.

"When are you arriving from, sir?" enquires the lady.

"Early twenty-first century," you reply, trying to sound as familiar with this kind of travel as is possible.

"Please follow me, sir," she says, her tone making it clear that this order should not be mistaken for a request.

As pretty much all the other arriving passengers nearby cast a reproachful glance at you for being in trouble, you step out of the line and follow the officer through the hall.

"Is something wrong?" you ask, as a door slides open in front of the customs officer.

"Please step inside, sir," is the only reply you get.

Inside, another (rather hostile-looking) officer is seated behind a large desk. Behind him, above his head, a large sign reads:

"Time-travel-distress psychological ward – Any offence against our staff will lead to immediate prosecution."

Obviously displeased to have yet another patient to deal with, the officer impatiently waves for you to sit down in front of him.

Looking around in despair, you begin to break out into a sweat. The room is empty. There's just the desk, the unfriendly officer, the sign and . . . and a now familiar yellow tube protruding from the side of the desk. All your worries are immediately blown away as you recognize your particle-throwing companion.

Is this another simulation? you wonder. If so, it has certainly made you feel a little better about your place in the universe, and reflective about the nature of life and death.

The quest to understand reality is a personal one, when all is said and done, and neither the supercomputer nor I should impose our views upon you. It's your right to have ideas of your own. Still, I should here warn you that, so far, you've just skimmed over the two theories that scientists use to describe our universe: quantum field theory and Einstein's theory of gravity.* They certainly both look coherent and elegant, but you should know that there are problems with many of the concepts they involve.

In fact, to be completely honest with you, *no one* really understands the universe yet. This is perhaps one of the reasons why we are all so happy when something that was predicted is found, like the Higgs boson or gravitational waves.

Even the reality that is around you, right now, or on your sofa, or on a tropical beach, is clouded in mystery. But one thing is for sure. All the mysteries that count, whether they be around you or within you, or far away beyond the Big Bang, eventually lead to

---

* Special relativity, Einstein's theory of (fast-) moving bodies, is included in them both.

the unification of the quantum fields with a quantum theory of gravity.

And while it is true to say that such a theory of Everything is not known, at least one of the properties of quantum gravity has been found. A clue, if you like. A clue that gives a tantalizing hint of what lies beyond the Planck wall.

That's the good news.

The bad news is that there's only one known window that opens onto this clue – a window that suggests it might one day be possible to travel, if only in our minds, beyond the origin of spacetime. And that is why the robot has come to collect you at the timeport. As the room you're in disappears to reveal yet again the dark landscape of deep space, you begin to ask about your destination but are cut off mid-sentence.

"I'm taking you to a black hole," announces the machine.

As you travelled to one right at the start of your cosmic adventures, you are perhaps wondering what you missed on that first visit.

The answer, for once, is rather simple.

You didn't get close enough.

Part Six

Unexpected Mysteries

1 | *The Universe*

When you think about it, there is something peculiar about the universe we belong to. Its name, *universe*, comes from *uni* ("one") and *verse* ("turned into"), so its name basically means "turned into one," highlighting, from the start, a very particular problem it raises.

Any experiment made *within* our universe can be repeated many times over. Want to check Newton's law of gravitation on Earth? Shoot an arrow. Not sure you got it right? Shoot another one. Again and again. With patience you will see that, from the knowledge of its initial position, angle and velocity, you can predict where the arrow will land. That is what ballistics is all about. And it works. Bows and arrows would have been abandoned long ago otherwise, and England would be French.

So, with a law and an initial condition, you can, it has been proved, predict where an arrow lands and defend a whole country.

For the universe as a whole, it is a bit more difficult.

Even if you had a law that explained everything, that applied everywhere, how would you make it work? How would you use it to predict how the universe we live in today became the way it is? You'd need an initial condition for that. Which you don't have.

But you could try to outsmart nature. Starting today, running time backwards, you could maybe reach an initial event that occurred a long time ago. That is what scientists have done. That is what *you* did back in Part Five. And they and you reached the Planck wall. Which is a pretty good start, since it corresponds to when space and time became what they today are.

But that doesn't remove the frustrating fact that, unlike with your experiment with the arrows, you only have one universe to play with. You can't have a go at creating another one, with different initial conditions, and check what comes out of it. Not in the lab, anyway.

But what if ours *isn't* the only universe? What if we were part of yet another type of multiverse, different from the one you were introduced to at the end of Part Two? Could *our* reality be only one of a myriad of possible realities, all having different beginnings, maybe even different laws, and hence very different presents?

The idea of such a multiverse is a question you will soon have to face, because it forms part of the answer modern theoretical physics has come up with to the mysteries you will probe throughout this part.

Indeed this section of the book is going to be a bit different from the previous ones. In Parts One and Two, you travelled through the very big. You learnt about gravity. In Part Three you saw what our reality looks like when travelling very fast and then, in Part Four, you entered the realm of the very small. In short, until now, you've probed the relativity of time and space, and quantum physics. But nowhere, until now, did you mix gravity and quantum ideas. This is what you will aim to do here.

For this, you will have to exercise your mind a little, just as you would exercise your body by stretching it.

Mixing gravity and quantum physics means mixing the very big and the very small. So, to ready yourself, your mind will have to learn how to jump from the very small to the very big, and back to the very small, again and again.

In doing so, you will see what goes *wrong* with the theories you've seen so far.

When that is done, you will journey with your robot guide to a place where gravity *and* quantum effects are both at work.

For now, however, let's have a look at the mysteries of modern science together, just you and me.

One can argue that there are three kinds of mysteries in physics.

The first ones are inherent to the theories themselves; they are theoretical. The second ones are rooted in observations and experiments. These are the ones that usually, but not always, drive research. The third type of mystery arises when no one understands *anything* any more. Black holes and pre-spacetime physics belong to all three types. They are both bridges and obstacles that lie between us and the holy grail of modern research: a theory that unifies the quantum world and the dynamical aspects of spacetime that Einstein figured out. That's why they're exciting.

And that's why the robot is keen to take you near to a black hole.

But why a black hole? Why not to the origins of the universe itself?

Because in the case of both the black hole and the birth of the universe, a huge amount of energy is confined within a pretty small volume. In both cases, the very big shrinks into the very small, and in both cases, neither gravity nor quantum effects can be ignored.

In that sense, black holes and our universe's origin look very similar.

One can't look at the universe from the outside though, of course. Experimentally, even if we had a law that governed the behaviour of everything there is, visible or not, we couldn't check whether different initial set-ups give different evolutionary models for our universe as a whole. We can't create Big Bangs in the lab, and we don't see new universes appear out there in the night sky, for us to analyse them.

Which is why black holes are useful.

For one, there are lots of them. Pick pretty much any galaxy in the universe, and there is probably a supermassive black hole at its centre. There may also be many more, smaller ones, with masses a few times that of our star, distributed everywhere. As of 2017, the largest black hole ever detected is 23 billion times the mass of the Sun. It lies some 12 billion light years away, in what was a very young galaxy at the time it emitted the light we catch today. At the other end of the scale, theoretically, the smallest black holes can measure anything down to the so-called *Planck scale* limit, which corresponds to an environment where gravitational and quantum effects both have to be taken into account. In numbers, the Planck length corresponds to 16 millionths of a billionth of a billionth of a billionth of a millimetre. So tiny that for all practical purposes, black holes can be of any size at all.

Both black holes and the very early universe share some important common features. They both involve a limit beyond which gravity cannot be used without incorporating quantum effects. That limit is the Planck wall, the wall you saw when travelling back in time beyond the Big Bang, at the end of the last part. Around the universe's birth, this wall was everywhere. For black holes, however, it is normally hidden from view, behind a gate that only opens one way: a *horizon*. You will cross one at the end of this part.

That trip will be the key that will lead you to Part Seven, where you will embark on your ultimate journey: a voyage through the universe as seen by the most popular of modern theories, a vision of Everything that endeavours to unify space, time and quantum fields. But these theories, called *string theories*, are so crazy, entailing both multiple and parallel universes, extra dimensions and all, that you could very well start believing that scientists have rather lost their marbles.

If it were not for the mysteries they solve, that is.

After all you've been through to reach this page, you may find it amusing to learn that, far from having discovered nearly everything, twentieth-century physics has left us with a picture of our universe that is mostly filled with deep and dark unknowns. This should not, however, come as a disappointment. These unknowns are the (opaque) windows we have on the science of tomorrow. And, between you and me, seeing how much mankind's understanding about pretty much everything has evolved in less than a century, seeing the baffling ideas that are today sprouting in the minds of theoretical physicists, there is little doubt that more revolutions of thought are still to come. Some may even be ripe or ready to bloom, just lacking the right experiment, ready to mould our perceptions with the promise of a strange, magical new reality.

So, here is what is going to happen to you now.

First you will have another look at the quantum fields that fill our universe, and you will see that, despite what I've told you so far, they make absolutely no sense. Then you'll have yet another look at all the particles these fields give rise to in the context of quantum gravity, and you'll see that they make no sense whatsoever either. Then you'll meet a cat, which is both dead and alive, and you will, if you follow it all, not understand anything any more.

Strengthened by these successes, you'll hear about parallel universes splitting off from ours like branches on a tree.

Once convinced that the quantum world is utterly beyond what our common sense would have us believe reality is like, you will move on to more familiar territory. Aiming, eventually, to bridge the gap that separates the very tiny and the very big, you'll turn to the big picture again, for a fresh look at Einstein's theory, at our universe's galaxies, at its expansion, hoping to find it all reassuringly well-defined. Funnily enough, you won't. You'll see for yourself that most of what our universe contains is not only invisible to our telescopes, but also unknown. The universe on a large scale is filled with mysteries wherever you look, just as it is on a very small scale.

Reluctantly or not, you'll then have to digest the fact that, however powerful it has been and will for ever remain, Einstein's theory of a curved spacetime is incomplete, that it even predicts its own breakdown and therefore cannot be a theory of Everything. There *are* places within our universe where it cannot be used. This means that a bigger theory needs to be found, if one ever intends to explain everything.

Where does it break down?

You've probably guessed: within black holes and before the Big Bang, somewhere on the way to the Planck wall.

So far, you have been travelling through the best theories mankind has ever built to describe the world that surrounds us. In practice, it means you now know as much about our universe as a good graduate student from any of the best universities on Earth. Not in technical terms, obviously, but certainly in terms of ideas. It should already be enough for you to shine at any dinner party.

Now it is time for you to go further and see what *doesn't* work. And then you'll not only shine, but also be able to make your friends scratch their heads in disbelief.

2 | Quantum Infinities

Do you remember seeing what the outer-space vacuum "really" looks like? What had until then seemed to be but emptiness turned into a wilderness of fluctuating fields. Fluctuations became particles that popped out of the fields' vacua, everywhere.

In the quantum world, when something is possible, it happens. So, forget about your normal everyday size and about gravity for a moment and picture your mini-self immersed within quantum fields, in the world of the very small, sitting on a mini-chair. You are like a referee, watching two electrons interact with each other, just as you might watch a game of tennis, the electrons being the players, the balls the virtual photons that dance between them.

There is an electron somewhere to your right, and another to your left. Being exactly identical, they both carry the same electric charge. Like magnets, they should repel each other. This should be fun to watch. The electrons are far away for now, propagating within the electromagnetic field they were born from. They move closer to each other, they are about to hit each other but they don't. They interact. They play. Virtual photons pop out of the electromagnetic field, deflecting the electrons, scattering them. Then, as quickly as it began, the game is over. The electrons and the virtual photons are gone.

You wait for the next game.

Another pair of electrons is on its way.

You decide to focus on the virtual photons this time, rather than the electrons. You sharpen the focus of your mini-eyes.

The electrons are moving. They are getting closer and closer and – pop! – virtual photons appear. Not to miss anything, you slow down the passage of time.

The electrons are about to be deflected.

The virtual photons are there all right.

~~But something is happening.~~

One of the virtual photons that appeared in between the two electron tennis-players spontaneously underwent some strange metamorphosis.

It became a particle–antiparticle pair: an electron and a positron.

You cast a quick glance towards the electrons, curious as to whether they might be affected by having lost their virtual pearl of light, but they don't seem to mind at all, so you look back at the pair just created and . . . it is not a pair any more, but two and a half.

You close your mini-eyes and rub them.

What kind of game is this?

You open your eyes again.

There are suddenly thousands of pairs of particles and antiparticles between the two electrons.

You blink.

There are hundreds of millions of them.

And now thousands of billions.

You blink again and . . . they're all gone.

You check the electrons.

They've scattered. Just like the previous players. Amazing.

What you've witnessed is one consequence of the quantum rules that apply to the very small: if something is possible, it happens. And it is very possible for virtual photons, dipping into the moving electrons' energy, to turn into virtual pairs of particle–antiparticles, which in turn can become further pairs of particle–antiparticles, which in turn can become further pairs, or annihilate back into light, which in turn can . . .

You get the idea.

Even when just two tiny electrons interact, the possibilities of virtual pairs appearing during the interaction are infinite. And so an infinite number of virtual pairs is involved.

Pondering this, still sitting happily atop your mini-referee's chair, you wait for the next game in order to witness the fireworks again, but there aren't any more players. No electron is coming your way. Still, now that you know what to look for, you see virtual pairs of particle–antiparticles appear all the same, albeit at a slower pace. They look like tennis balls and antiballs popping out of nowhere, with no players around.

These pair-creations are the *quantum fluctuations* of the vacuum.

They are present all the time, but when there is some available energy for them to tap into – like the kinetic energy of some incoming electrons – they get much more excited.

An electron–positron pair spontaneously appears in front of you and the pair annihilates into a photon which spontaneously turns into another pair, a quark–antiquark pair, and now one of the antiquarks emits a gluon, which in turn . . .

Even in the vacuum, when there seems to be nothing around, to build a correct picture of our world, all the infinite possibilities of particle–antiparticle creation have to be taken into account, everywhere, all the time.

A mess.

A mess with a rather catastrophic consequence: the possibilities are so important and numerous (they are in fact infinite) that there should be an infinite amount of energy at each and every point of our universe. Even where there is nothing else, in the vacuum. Rather obviously, this is not the case, or our universe would collapse everywhere, right now, because of the extraordinary gravitational impact this would have on spacetime. So there's something wrong with this picture.

To make this rather cumbersome problem easier, quantum-field theorists came up with a pretty cunning trick: they purely and simply decided to forget about gravity, to entirely remove it from the game. And, while they were at it, they got rid of the infinities too. They turned them off and made calculations with what was left and, abracadabra! . . . it worked.

Dutch theoretical physicist Gerard 't Hooft, to name but one of the amazing and brilliant physicists who fathered such mathematical surgery, jointly received, with his PhD supervisor Martinus Veltman, the 1999 Nobel Prize in Physics for this. Thanks to them (and a few others), and despite the mathematical hocus-pocus to account for the infinities, quantum field theory arguably became, through its predictive power, the most successful scientific theory of all time. Getting rid of the infinities led to predictions about particles that hadn't even been seen before, predictions that were accurate – as far as their mass or charge was concerned – up to one part in more than 100 billion. Were a randomly chosen person to be that precise, he or she would be able to guess if a drop of beer was missing from a million pints served in a pub. Riots would no doubt occur on a daily basis if we had such a capacity.

Quantum field theories are astonishingly precise in their predictive power, but this trick leaves us all frustrated for reasons that even a million beers can't help wash away.

Why do these infinities occur?

Could it just be because we do not know what is happening in regions of our universe that are even smaller than the ones these theories are probing?

Perhaps.

That is what one extraordinary American physicist thought, anyway. His name was Kenneth Geddes Wilson, and instead of trying to explain infinitely smaller realms to reach a conclusion about particles, he thought that such dizzying scales might actually be the problem: that one should not necessarily have to consider ever-smaller scales in order to be able to talk about particles. Just as one does not need to know about atoms to compare apples on a market stand, Wilson argued – and proved – that what is not known could be gauged and codified and forgotten about.

And it worked – indeed he received the 1982 Nobel Prize in Physics for it.

Wilson didn't solve the problem of what happens in the infinitely small, though, he just got rid of it. Introducing a cut-off, coarse-graining what wasn't known, the infinities that had marred the field before did not occur any more.

The process of removing infinities has a name. It is called *renormalization*. As I said above, it is brilliantly efficient for making calculations. But to ever have a hope of understanding everything, one cannot just bypass the unknown. One needs to dive into it. Especially so because with gravity, these renormalization procedures do not work.

Quantum field theories are about what the universe contains. They are very accurate, astonishingly accurate really, but only when they leave the background spacetime alone, fixed, with gravity having no effect on anything whatsoever. Not a very realistic world.

We need to find a way to bring gravity back in.

We need to turn gravity into a quantum field.

So how can that be done?

Quantum field theories assert that as soon as there are fields around, these fields can create small packets of energy, or small packets of matter, which are called *quanta*.* The basic quanta of the electromagnetic field are the least energetic states of its elementary

---

* The word "quantum" literally means "small packet," from the Latin, and *quanta* is the plural.

particles, the photons and the electrons. Similarly, the basic quanta of the strong nuclear force field give the quarks and the gluons, whereas the basic quanta of the gravitational field, considered as a hypothetical quantum field, are what we earlier called the gravitons.

You heard about these before, in Part Five, but we dismissed them back then. Why have they reappeared here? Because we'd like to see what could be wrong with them.

So let's think of gravity as arising from a quantum field like all the other fields you've seen so far. The gravitons then are its force carriers. And when they calculate, on paper, how these gravitons would affect their surroundings, theoreticians find that they would be just like the curves of spacetime.

On paper, they *are* gravity.

A very promising start.

But it so happens that when they think further, scientists realize that those very quanta of the gravitational field, those gravitons, also make the straightforward idea of gravity fail completely.

Which is not a very good thing.

Why is it so?

First, gravitons have no reason not to interact with one another: were they to exist, then they should, by all means, be as subjected to gravity as everything else, and hence to themselves.

And second, being elementary particles of a quantum field, they should also be able to appear everywhere out of their field's vacuum, leading to infinities just like the ones that 't Hooft and Veltman cured. This time, however, the gravitational quantum infinities cannot be removed by any renormalization procedure: here, 't Hooft and Veltman's machinery fails utterly, and Wilson's approach doesn't apply at all, since it ignores the very distances at which the gravitons act.

All in all, this means that really problematic infinities arise when trying to turn gravity into a quantum field in a standard way, and one obviously cannot turn a blind eye to gravity to get rid of them, because gravitons *are* gravity.

If gravity was a quantum field as we just mentioned, if gravitons were a correct description of how gravity works in nature, spacetime should react to these infinities and collapse pretty much everywhere. Which it doesn't. Or we wouldn't be here to talk about it.

Funnily enough, despite all this, and you may believe they are lunatics, many scientists (including myself – I'll show you why in Part Seven) believe that gravitons do exist, at least as part of the bigger theory that everybody is looking for.

Now, while we're at it, let's go further still, so that you see, right from the start, several reasons why Einstein's theory of general relativity and quantum field theory are at odds.

Gravity has to do with spacetime. With space and time, that is. Intertwined.

In a quantum field theory, the elementary particles that pop out of the vacuum are made of the field itself. In the case of a quantum field theory of *gravity*, the elementary particles should therefore also be made of their field. But that field is spacetime. So the particles should be made of space and time themselves.

This means that there should be fundamental packets of spacetime around, everywhere, and, incidentally, that neither space nor time should be continuous.

Worse, these packets of spacetime should be able to behave both as waves and as particles. And be subjected to quantum tunnelling, to quantum jumps . . .

Good luck trying to picture that in your mind.

In fact, if you are a normal human being, just trying to think about it should make your brain melt.

As far as nature is concerned, however, it should not be a problem.

A real problem, though, is that even if we forget about the troublesome infinities, all the other quantum field theories that are so powerful in describing all the particles we are made of only work as long as there are no such packets of spacetime around.

In other words, it means that general relativity theory and quantum field theory do not use the same notions of space and time.

And that is a problem.

A very big problem. With no obvious solution.

And one is therefore left with a curious feeling of being stuck in the middle: mankind has uncovered two extremely efficient theories: one that describes our universe's structure (Einstein's gravity: the theory of general relativity), and one that describes everything our universe contains (quantum field theory), and these two theories won't talk to each other. For a very long time, even the physicists working on each of these two fields followed suit and did not talk to each other either. US theoretical physicist Richard Feynman, Nobel Prize laureate for his work on quantum field theory and one of the most brilliant scientists of all time, wrote a classic letter to his wife accounting for this: "I am not getting anything out of the meeting," he stated, in 1962, after attending a gravity conference. "I am learning nothing. Because there are no experiments, this field is not an active one, so few of the best men are doing work in it. The result is that there are hosts of dopes here (126) and it is not good for my blood pressure. Remind me not to come to any more gravity conferences!"

Yet, thanks to new technologies and the work of theoretical physicists such as Stephen Hawking, scientists soon found out they could not turn a blind eye to what they did not know, and ideas from both sides began to spill over to the other, giving birth to the crazy ideas you will travel through in Part Seven, to which I'll now introduce you.

Do you remember those quantum particles the robot played with in the white room with the metal post? Down there, in the world of the very small, particles really do take all possible and impossible paths to get from one place to another, from one time to another, as long as no one looks.

So why don't all the quantum aspects of all the particles that make up your body turn you into a quantum-you?

Wouldn't that be cool?

All the different life-paths you could possibly imagine would simultaneously occur. You'd be very rich and very poor and married and single and happy and sad and win a Nobel Prize and be completely dumb and be here and there, and live now and then . . . You'd live all the lives you could dream of, really, and all the ones you wouldn't want to.

But this doesn't seem to be happening.

You are made of quantum stuff, aren't you? So it should happen.

But it doesn't.

Why?

Well, as amazing as it may sound, no one knows. In fact, it is linked to one of the greatest mysteries of the quantum world: why do we not see quantum effects everywhere around us?

Being made out of quantum particles, expressions of quantum fields, like everything else, why do we experience the world the way we do and not the way particles do at a tiny, subatomic level?

One could argue that the world is the way it is, that physics is not about questioning its rules but about trying to decipher them.

There is, however, a slight problem with such a humble statement: the laws of the quantum world are so very different from the reality we perceive on a daily basis that there should be some sort of transition between the quantum world and the *classical* one, as the one we experience, the one we are used to, is called. Were the particles that make up our bodies, or that are found in the air or in outer space, to behave like decent tennis balls or baseballs, then everything would be fine. We would understand everything from the tiniest elements to the largest ones.

But they don't behave like that.

You've seen they don't many times throughout your journeys in the world of the very small. When trying to catch the electron that swirls around a hydrogen atom, for instance, do you remember how hard it was for you to know both where it was and how fast it was moving? Well, let's have another look at this fact now.

Picture yourself in your mini-you state. You are smaller than an atom. A particle is on its way towards you. You know nothing about it, neither its size, nor where it is, nor how fast it is approaching. You just know it obeys the rules of the quantum world.

You take out a mini-torch from a mini-bag you brought along and get ready to switch it on, expecting its light to bounce off the particle, wherever it is, and travel back to you, thereby telling you its position.

But you can't use just any light to do so.

You need to use the "right" light.

Remember that light can be thought of as a wave? Well, the "right" light here means that the separation between two consecutive wave-crests (its wavelength) has to be roughly the size of your target or smaller. Use too large a wavelength, and the light it corresponds to won't notice the particle at all. It will shoot through it, just as radio waves travel through the walls of your house without even noticing them. With the "right" wavelength, however, you'll get a bounce-back and you will be able to tell your particle's position with the accuracy of the wavelength used. Simultaneously, you'll be able to check what the particle's velocity is and you'll know everything you want to know about it.

Easy.

You adjust the knob of your state-of-the art mini-torch to get a very energetic pulse. Focused, you shoot and . . . *Bam!* You hit something. A particle. There. Ahead of you. Light bounced on it and travelled back to you. The time it took to travel there and back tells you exactly where the particle was at impact, and the particle therefore cannot be everywhere any more. Once detected, the particle loses its quantum wave attributes. From all the possible positions it held simultaneously a fraction of a second before, one was picked up by the very act of using your torch as a probe. Just like when the robot threw that particle in the white room, the particle went everywhere *until* it got detected by a detector. This irreversible process is called a *quantum wave collapse*.

After the collapse takes place, you know where the particle is, up to the accuracy of a wavelength. Now you want to know how fast it was moving at the point of impact.

But that is not going to be so easy.

In fact, you won't be able to answer that question exactly.

Ever.

Remember: the shorter the wavelength, the more energetic the light it corresponds to.

So the more precise the position you get, the more energetic the light you had to use for your torch, the harder you hit your particle – and the less you therefore know about its subsequent velocity.

In the world we are used to, this is a trivial statement.

Try, in the dark, to pinpoint the position of a moving object by firing something at it. The impact will affect what you wanted to probe. If what you fired bounces back to you, you'll know where the object was at impact, but if you fire once more to know where it went, you'll see that its velocity changed because of your first shot.

Trivial indeed.

In the quantum world, however, that is *not* just a trivial uncertainty. It is a deep property of nature. It says that you fundamentally cannot know both where a particle is and how fast it is moving. This rule is called the *Heisenberg uncertainty principle* after German theoretical physicist Werner Heisenberg, who discovered it. Heisenberg is one of the founding fathers of the quantum theory of the atomic world. He received the 1932 Nobel Prize in Physics for it. He knew what he was talking about. But, like everyone else ever since, he did not *understand* it. It is beyond our intuition, it is contrary to common sense.

The uncertainty principle immediately makes the quantum world very different from our everyday, classical world.

Right now, with respect to your body, you know where the book you are reading is *and* how fast it is moving. You hence know its position *and* velocity with a pretty good level of accuracy. Still, there is an uncertainty about its position and velocity too – an uncertainty too small for you to notice, and therefore an uncertainty that doesn't really matter.

In the very small, however, in your mini-you state, you wouldn't be able to hold a book in your hands, or even a torch. Were you to precisely know where a mini-copy of this book was, the uncertainty about its velocity would be huge because you'd be firing many particles at it just to know where it was, and you would never be able to look at it. Alternatively, were you to know precisely how fast it was moving, you wouldn't be able, by any means, to know where it was, making it tricky to read. In the very small, position and velocity merge into a hazy concept. Just like the Casimir effect, as technology becomes smaller and smaller, this is again a problem that engineers will increasingly have to face.

That said, the Heisenberg uncertainty principle is not a mystery. It is a fact.

Strictly speaking, it isn't even an uncertainty. It just says that our classical notions of position and velocity do not apply in the very small. Nature works in a different way down there, and we do have theories to account for it, theories that predict it: quantum physics. And these strange effects *do* carry through to our scales, but we're just not built to feel them. They become insignificant when too many particles are involved. And that is a well-understood fact too.

So where is the mystery we were looking for? Is there one?

There is.

We overlooked something in the measurement you've just made: the quantum wave collapse itself.

And that *is* a mystery.

A really puzzling one.

When left alone, quantum particles behave as multiple images of themselves (as waves, really), simultaneously moving through all possible paths in space and time.

Now, again, why do we not experience this multitude around ourselves? Is it because we are probing things around us all the time? Why do all experiments that involve, say, the position of a particle make the particle suddenly be *somewhere* rather than everywhere?

No one knows.

Before you probe it, a particle is a wave of possibilities. After you've probed it, it is somewhere, and subsequently it is somewhere for ever, rather than everywhere again.

Strange, that.

Nothing, within the laws of quantum physics, allows for such a collapse to happen. It is an experimental mystery *and* a theoretical one.

Quantum physics stipulates that whenever something is there, it can transform into something else, of course, but it cannot disappear. And since quantum physics allows for multiple possibilities simultaneously, these possibilities should then keep existing, even after a measurement is made. But they don't. Every possibility but one vanishes. We do not see any of the others around us. We live in a classical world, where everything is based on quantum laws but nothing resembles the quantum world.

So the question is: how can we make quantum effects appear at our human scale, so that we can probe them and see the collapse, if there really is any, with our very eyes? Is that possible? And if one could see quantum effects like this, what would we expect to see?

In 1935, two years after being awarded the Nobel Prize for his work on quantum physics, Austrian physicist Erwin Schrödinger devised an experiment to bring quantum effects to our scale. It involved a cat, and a box. Even though it was only ever a thought experiment, no scientist has stopped wondering whether that cat is dead or alive ever since.

You are about to do Schrödinger's experiment again. And I hope you're not too fond of cute, purring, innocent, playful kittens: there is a good chance the cat will be harmed during the experiment. In any case, do please keep in mind that the idea here is to make quantum effects appear macroscopic. Some sacrifices may be necessary.

With that disclaimer behind us, let's start.

For those of you who do not know, a cat is a four-legged, usually furry, tailed mammal which lives within the same scales of reality as we do. Most people like to cuddle them, but not all. They come in pretty much all colours but, as far as I know, not in green.

To perform Schrödinger's thought experiment, you decide to pick an adorable little kitten, black and white, and you look for a box that can be so perfectly sealed that, once closed, no one can know anything about its interior from the outside.

Besides the cat and the box, you need to fetch a radioactive substance, a very special one that is known to have a 50 per cent chance of emitting some radiation during the time of your experiment. Radioactive materials are very unpredictable. According to quantum laws, there is no way whatsoever to know in advance whether they will decay and emit some radiation or not. There is just a probability. One chance out of two, for the substance you've found.

Now, you also need to find three other objects: a radiation detector, a hammer, and a vial containing some very lethal poison.

Then you link everything together so that if the detector notices some radiation emitted by the radioactive substance, the hammer breaks the vial and the poison is released. This would be harmless, were it not for the fact that you put all of these things, the hammer and the radioactive substance and the poison and the cat, in the box – and seal the lid.

Then you wait.

What then?

There is a 50 per cent chance that the cat gets poisoned. Everything depends on the radioactive decay.

A twisted experiment, agreed.

You definitely should not try it at home.

Now, here comes the question: is the cat dead?

Quantum effects are in progress here, as desired. And the outcome is macroscopic – big enough for us to see.

But without opening the box, there's no way for you to know whether the radioactive decay occurred or not, so there's no way you can tell whether the vial is broken or not and hence whether the cat is dead or alive.

Nothing new under the Sun, you think? Well, with all things quantum, one ought to be vigilant and use common sense sparingly. Or indeed not at all. To infer anything down there, one needs to abide by the laws of the quantum world. In real life, one could expect the cat in the box to be dead or alive.

But then both those answers would be wrong.

In the quantum world, what can happen happens. You should be used to that by now.

Here, the decay *and* the non-decay of the radioactive substance can happen with the same odds, so both *do* happen. Just as a particle can travel to the left *and* to the right of a solid post simultaneously, radioactive decay also simultaneously happens *and* doesn't happen, as long as no one looks. As said above, most of the time such a *superposition* of possibilities goes unnoticed by us because, for some unclear reason, it never occurs at – or reaches – our scale. In our particular experiment, however, the set-up is designed so that our eyes can see it: the simultaneity of two quantum possibilities (decay and non-decay) is directly linked to the rather dramatic death or survival of a cat.

So what do the rules of the quantum world say?

They say this: the decay and non-decay events being directly linked to the poison, the cat, as long as the box is not opened, should be neither dead nor alive, but both.

Before you open the box, the decay has both occurred and not occurred, so the poison has been released and not released.

So the cat is dead and not dead.

Dead *and* alive.

Hearing this, you immediately open the box to check it out.

The cat jumps out, unharmed, and so very cute.

And there's no carcass lying at the bottom.

You scratch your head.

This whole business of "superposition of states" and "subsequent collapse of quantum possibilities" suddenly sounds like a pretty elaborate trick rather than a real phenomenon.

Did we get it wrong? Was the cat really dead *and* alive for a while or was it all a cheat?

Let's see.

Opening the box made you interact with the experiment, right? Aha.

So you did interfere. You *did* look. And when one looks, nature has to choose.

So the choice, the collapse, if real, must have happened, leaving the cat alive.*

But did the cat's fate freeze before you opened the box? Or did it do so after, extremely quickly?

You are back to the initial question: did a collapse happen at all?

Schrödinger devised this thought experiment in 1935 and, for years, no one could answer his puzzle, until French physicist Serge Haroche and US physicist David J. Wineland managed to devise a real experiment able to detect the very superpositions that were then supposed to collapse.

They did not use a cat, though.

They used atoms and light.

And they saw that quantum superpositions are very real; that pretty much any quantum particle can, and does, simultaneously exist in different, mutually exclusive states. Today, this is actually the basic reason why engineers are trying to build quantum computers. Using the ability of quantum particles to be in different states at once, computers can, in principle, become exponentially more powerful than what can be achieved with our classical ones, making "parallel" computations simultaneously. Haroche and Wineland jointly received the 2012 Nobel Prize in Physics for this. They somehow proved that Schrödinger's cat really was both dead and alive, simultaneously, at some stage.

Now where is the mystery here?

It has to do with what is gone.

---

* It could as well have been dead, but happy endings are more popular.

The superpositions are real, fine. That's what Haroche and Wineland proved. We have to accept it.

But when you opened the box, when the collapse occurred and the live cat jumped out, where did the possibilities you did not see go? Since it must have been real at some stage, where did the dead cat go?

That is the mystery.

Many scientists have wondered about this, and a few putative answers have recently begun to flourish. Some guess that the possibilities that were not observed fade away, like drops of ink within a lake, the lake here being the world we live in, as if pearls of unfulfilled possible realities diffused within the one and only reality that prevails, the one we are part of. Some others believe that our consciousness has something to do with it all, that it is the very act of experimenting or even thinking that freezes reality in one state, thereby creating it.

And then there is US theoretical physicist Hugh Everett III.

Born in 1930, Everett was a very strange man. Extremely bright, he studied mathematics and chemistry and physics to eventually write a PhD thesis under the supervision of one of the most influential US physicists ever, John Archibald Wheeler, from Princeton University. Everett gave up physics right after his PhD, though, mostly because he apparently believed it was all too weird. Wheeler's failed attempts to have his student's ideas taken seriously by the scientific community probably didn't help either. At twenty-one, leaving theoretical matters behind, Everett began to work for the top-secret weapons branch of the United States military and eventually died of too much drink and too many cigarettes. In an uncanny resemblance to the lives of some famous poets or painters who silently burn their talent in their early years while held in contempt by their peers, Everett's 1956 dissertation later became a classic. In it, he made the extraordinary claim that since quantum ideas were working so well at very small scales, they should be taken seriously all the way through to our scale. Everything in our universe being made of quantum stuff, everything should therefore be thought of as a huge quantum wave of possibilities existing simultaneously.

With such a point of view, no collapse then ever occurs. Every possibility exists.

With such a point of view, the whole universe branches off every time a choice has to be made as a result of an experiment or anything else. Unfathomably many parallel universes should therefore exist, where all the possibilities, all the alternative outcomes, are facts.

According to Everett, parallel histories should surround us.

You hesitate between two elevators before picking one? Another you, in a branching parallel universe, chooses the other. In another one, you hit the wall in between the two. In yet another one, you take the stairs. All possibilities are thus fulfilled.

In a way, Everett's literal understanding of quantum physics says that if you put selfishness aside, you should never be sad. Whenever something bad happens to you here, infinitely many parallel yous in infinitely many parallel universes escape the bad news and feel happy.

Everett is still alive and even reading this book in yet another infinity of these parallel realities. In some he likes what I write about him. In others, he doesn't. In others yet, he actually wrote this book himself and, in it, Schrödinger's cat is a green dog.

According to Everett's interpretation, no real choice is ever made by nature. All possibilities happen.

You just do not know about it.

No wonder he gave up physics.

Everett's idea is weird indeed, but it has now been taken seriously by some of the greatest physicists of our times, with many mathematical models pertaining to the origins of spacetime making use of it. There certainly is no experimental confirmation (or rejection) in sight for Everett's claim, but it is an appealing reason why the reality we live in is not a superposition of quantum possibilities: the possibilities we do not experience are real enough, but elsewhere.

Now, while you get used to this idea, let's quickly summarize what you've been through up to here.

Since the beginning of your journey, you have been travelling within the very big and the very small separately. Shooting through cosmic kingdoms, you have discovered what the large-scale structure of our universe looks like and how it is ruled by general relativity. In the realm of the tiny, you've seen that the quantum rules of nature are different from the ones we are used to in our everyday business. Until this part, you have therefore been travelling throughout what is *known*, both theoretically and experimentally. You've seen what the universe appears to be, whatever the scale, from the point of view of an early twenty-first-century scientist.

In this part, you've started to glimpse the limits of this knowledge. You've seen that not only are general relativity theory and quantum field theory reluctant to talk to each other, the quantum laws do not appear to rule our daily lives, for reasons that may, for some, go as far as involving the existence of parallel worlds.

You'll see even weirder things in Part Seven.

For now, however, let's continue to exercise your mind and leave the very small to get back to Einstein. What about his theory? What are the mysteries that can be found there?

Are there any?

Are they as ubiquitous as the infinities that mar quantum field theory?

To both these last two questions, the answer is yes.

Forget about cats and dogs and parallel universes of alternative realities.

Forget about the quantum world.

Forget about your mini-you.

You are now in space, as a mind.

You've seen that the very small is filled with mysteries and you intend to check if Einstein's theory works everywhere, or if it too has shortcomings even without trying to turn it into a quantum theory.

So you are in space. The Earth is behind you and you fly forward. You pass the Moon, the Sun, the neighbouring stars.

Up to here, Einstein's theory of gravity works perfectly. Stars and planets move as they should.

You head out of the Milky Way, into the intergalactic medium, where you stop.

The Milky Way is beneath you, just there. Other galaxies shine in the distance. Huge spirals with hundreds of billions of stars that radiate light in an otherwise dark universe.

From what you've learnt about gravity, you know that, just like the planets around the sun, the velocities of any star within a galaxy can't be random. Stars that fly too fast should escape their galaxies' shelter and for ever wander, lone shiners, through the immense distances that separate galaxies from one another. Were the stars to move too slowly, they'd fall down the spacetime slope created by all the other stars, a slope that would effectively make them move towards the galaxies' core, the central bulge full of stars, where they'd end up being swallowed or destroyed by the giant black hole that patiently preys there. Without the right velocity for it to stay in a stable orbit, a star is either out of its galaxy, or it is doomed to fall, just as a marble spinning in a salad bowl either falls down or is ejected.

You remember that Newton's gravity failed when gravity was too strong. Near the Sun, his equations required corrections to account for Mercury's drift. Einstein found these corrections by revolutionizing our vision of space and time. And now, 100 years later, it is Einstein's turn to face a change of scale. What about Einstein's gravity around whole galaxies? Does his theory of spacetime curves work when faced with hundreds of billions of stars rather than just one?

That is what you are about to check.

You take out a stopwatch and start timing the stars as they travel throughout the Milky Way. To survey 300 billion of them at once is tricky, so you start with the outskirts, around the tip of one of the magnificent spiral arms, far away from Sagittarius A*, our very own supermassive black hole.

You count ten seconds.

The star you were timing travelled 1,500 miles. Not bad.

That corresponds to an average velocity of about 560,000 miles per hour around the centre of the galaxy. Not bad at all.

Its neighbouring stars are just as fast.

Any two stars that are the same distance away from our galaxy's core have the same velocity. Slower ones should be lying further out, while the fastest, like the speedy S2 you encountered a while ago, should lie deep within. And if you wonder how long it takes one of the outsiders to complete an entire run around the Milky Way, the answer is . . . about 250 million Earth years. A long journey. The Milky Way is big. The Sun (and therefore the Earth), being a bit deeper inside, travels around the Milky Way in slightly less than 225 million years, a period called a *galactic year*. The last time the Earth was at the galactic position it has today, the dinosaurs still had 160 million years to live . . . Using such terminology, the Big Bang occurred about sixty-one galactic years ago, and if we start from today, after twenty more rounds the Milky Way and the Andromeda Galaxy will be so close to each other that they will start to collide. Incidentally, the Sun will explode a few galactic months later. Put like that, it doesn't sound that far off . . .

All right.

So far, so good.

There doesn't seem to be any problem here with Einstein's theory, except . . .

Except that there is.

To be honest with you, you are not the first one to check out how fast these stars shoot around our galaxy. Their velocities have been known for quite a while, since around the early 1930s, when Dutch astronomer Jan Oort measured them.

But Jan Oort went a bit further.

First he estimated the amount of matter contained within the entire Milky Way. Then he checked whether the velocities he observed agreed with what was expected, for the stars not to fall in or shoot away.

They did not agree.

They did not agree at all.

Being up there, above the Milky Way, you can check it for yourself.

Adding the mass of every star and dust cloud and everything else you see belonging to our home galaxy, you reach the same puzzling conclusion: there is definitely not enough matter to keep *any* star from shooting away, given their velocity.

And unlike the mismatch between Newton's theory and Mercury's orbit, the gap here is not tiny at all.

There should be *five times* more matter than what you can see. Otherwise, the stars should all be shooting away. Including the Sun.

You must have overlooked something. And so must Oort.

It's not just a few hundred million stars and their dust equivalents that are missing, or you could have blamed yourself, or Oort, for poorly estimating the whole thing. That would have been acceptable, by the way. But five times the amount? What is going on? And who was this Oort in the first place? Can we trust him?

We can. He was not just a common astronomer. In fact, his incredible insights helped mankind figure out a lot of what you saw while travelling through the solar system and the Milky Way in the first part of this book. He is, for instance, credited with showing that the Sun is not at the centre of our galaxy (it may sound obvious now, but it wasn't before he proved it). He is also the man who hypothesized the existence of the huge reservoir of comets (billions upon billions of them) that now bears his name, the Oort cloud, which you crossed at the outer edges of the solar system before entering the gravitational realm of the red dwarf Proxima Centauri.

Oort was no ordinary scientist, then, and in 1932, to explain the absurd mismatch between the matter he could see throughout our galaxy and the velocity of its stars, he made an amazingly bold assertion. He stated that an unknown type of matter filled the Milky Way. A type of matter that had not yet been detected in any form whatsoever, neither here on Earth nor anywhere else, because it did not interact with light, making it impossible for anyone to see it with any type of light-gathering telescope. He called it *dark matter*. According to Oort, dark matter's visible effects are only indirect, through gravity: dark matter cannot be seen, but it bends spacetime like ordinary matter, even though ordinary matter it most certainly is not. It cannot be made out of the same particles that make up everything we know, otherwise we could see it.

Such a discovery may sound too big – and exciting – to be true, and however good Oort might have been, no one is perfect. He might have made a mistake. To check it out, you decide to look at other galaxies to see how they move around each other, just as Swiss astronomer Fritz Zwicky did about a year after Oort's initial claim, in 1933.

Were dark matter to be real and present and gravitationally active not just within the Milky Way but also inside and around other galaxies too, it wouldn't just modify the way stars move within the galaxies. It would also affect the way galaxies move around each other.

So you stare at them, focused.

You analyse the spectacular cosmic dance of these immense gatherings of shining stars and . . . you can doubt no more.

Just like Zwicky, you are forced to admit that all the galaxies move far too fast around each other not to hide an enormous amount of gravitationally attracting dark matter.

And dark matter is not matter.

It is not antimatter.

It is something else.

No one knows what.

Many other tests have been done since the 1930s and they have all reached the same conclusion. Dark matter is there. It exists. Everywhere there is matter there is dark matter smeared around. And even though I've tried, throughout this book, to *show* you everything I wished to share with you about our universe, at this point I have to admit that I can't bring you any closer to it.

Why?

Because even today, more than eighty years after Oort's bold guess, we still don't have a clue what this dark matter is made of. We know it exists. We know *where* it is. We have maps of its presence within and around galaxies throughout the universe. We even have stringent constraints on what it is *not*, but we have no clue *what* it is. And yes, its presence is overwhelming: for every one pound of ordinary matter made out of neutrons and protons and electrons, there are five pounds of dark matter, made out of who-knows-what.

Dark matter.

Unexpected Gravitational Mystery Number One.

It could mean that Einstein's theory doesn't work at such scales, just like Newton's didn't work too close to the Sun. But many independent checks have been made. Dark matter does indeed seem to be everywhere, around galaxies, around our very own Milky Way and throughout the universe, and you can't see it.

There seem to be many more invisibles in our universe than there are visibles.

Over the eons that passed after the Dark Ages of our universe, ~~many galactic collisions occurred, entire galaxies clashing and~~ merging. Violence, in outer space, is everywhere, and the galaxies you are now looking at are just the visible part of it.

Dark matter, overwhelming matter five to one, cannot be seen, and yet there is so much of it that it must have played – and still be playing – a role in this cosmic waltz you are witnessing. A waltz whose dancers, you now know, are gatherings of stars wrapped up in invisible cloaks made of dark matter.

The more you watch all these galaxies move around – the more dancers and shapes you see – the more worlds you start imagining out there, with skies completely different from ours. And you suddenly start wondering if some faraway civilization has already found answers to our human queries . . . But you freeze, blinded.

A very powerful source of light just hit your eyes.

You peer into the night to spot where it came from, but it is gone.

Just as suddenly, though, another one hits you, coming from some other, unimaginably distant place.

And another still.

Snapping out of your reverie, you focus your thoughts on the galaxies these light signals seem to originate from.

Without you knowing why, your heart is racing like mad. You look at their light, at the way they recede into the distance while moving around one another.

Something is wrong.

The galaxies these lights are coming from are not receding the way they should.

We're not talking about how they swirl around one another here, but about the expansion of the universe, about how *all* galaxies recede into the distance, like poppy seeds in a rising cake. Considering what you've learnt about this expansion, these galaxies are not moving right.

This is Unexpected Gravitational Mystery Number Two. And it involves much, much more hidden energy than dark matter.

To understand it, you need to know how to estimate distances in this universe of ours.

When you were lying on your tropical-island beach, just before you started your journey into outer space, how could you tell which star in the night sky was close, and which one was far away? Brightness obviously isn't enough. Stars come in pretty much all sizes and their actual luminosity can differ a great deal. A bright star, as seen from Earth, could be huge and very far away, or rather small but much closer. Some other trick must be used and, historically, scientists have so far used three different tools to estimate cosmic distances.

The first one concerns any object, be it a planet or a star, that is quite close to us. It is the easiest one of them all and uses common sense (there's nothing quantum going on here, so common sense is allowed). Picture yourself looking at a tree through the side window of a car on the motorway. Trees close to the road pass quickly by, while ones that are further away seem to move at a much slower pace. Mountain ranges, above the horizon, sometimes seem not to move at all. They can be used as a fixed background. In space, the same concept applies. As the Earth orbits the Sun, objects that are close by have an apparent motion that is rather obvious against the background of very distant stars, which seem fixed. Checking how much an object's position changes against that background while the Earth shoots around the Sun allows scientists to determine how far away in space that object is. It involves mathematics that Euclid would have understood, more than 2,200 years ago. It works very well for short-range estimates – within the Milky Way, that is. But it doesn't work for determining *galactic* distances. Galaxies are just too far away. As you, on Earth, orbit the Sun, your perspective on the cosmos can change by as much as 186 million miles from summer to winter, but that is not enough to see them move: galaxies are still part of the fixed background. To guess where they are, you need trick number two, which involves a very particular type of star called the *Cepheids*.

Cepheids are very bright stars whose lights oscillate between a maximum and a minimum level of intensity with impressive regularity. Amazingly enough, scientists have figured out a way to link this oscillating period to the total amount of light they shine. And that is all they need to tell how far away they are: just as the sound of a horn dims with the distance it travels from its source, so does light. Harvesting the portion of light emitted by a distant Cepheid as it reaches the Earth hence gives its distance away. And, rather fortunately, there are many Cepheids out there.

But this trick has its limits too: to measure the greatest distances in the universe, individual Cepheid stars cannot be used any more because even the most powerful telescopes cannot tell them from their surrounding groups of stars. To probe the very deep universe, a third trick is needed.

You may remember, from Part Two of this book, the work of US astronomer Edwin Hubble. In the 1920s, Hubble became the first person to observe that distant galaxies were all receding from us, that the universe is expanding. Some of your friends kindly confirmed that fact by making night-sky observations all around the Earth with their billion-dollar telescopes.

In the 1920s, Hubble used the colour shift of lights coming from far away galaxies' Cepheids to figure out their velocity and he observed that their eagerness to move away from us was proportional to their distance: a galaxy twice as far away as another moves away twice as fast. This law is now called *Hubble's law*.

Trick number three involves using Hubble's law the other way round, where Cepheids cannot be singled out from their surroundings. From the way colours are shifted in the light coming from faraway galaxies, scientists can tell how much of our universe's expansion these lights have travelled through. With this, it is possible to tell how far away the galaxy now is.

Hubble's law is simple enough and it fits rather well with what is known: space and time became what they now are some billions of years ago, spacetime has expanded ever since and, as seems normal for an expansion triggered by a violent release of energy (that's the Big Bang), the expansion rate slowed down over the billions of years that followed.

Everything is fine with this rather logical set-up.

Except that it doesn't fit with what you just saw.

The bursts of light your eyes caught are at odds with it. The way their colours were shifted does not match the grand, beautiful, coherent picture described above. Something is wrong, and mystery number two lingers there, somewhere.

To figure out what it is about, let's travel a little and have a look at what triggered the extraordinarily powerful bursts of light that hit your eyes.

Starting from above the Milky Way, you head for a particularly beautiful and colourful spiral galaxy lying about 8 billion light years away. You cross the immense, expanding distances that separate our cosmic family from this other island of lights and, once next to it, you enter it from the side. You fly by millions of its stars, shoot through dust clouds the size of thousands of solar systems combined, and suddenly you stop again.

Right in front of you are not one but two shining objects that attract your attention. They orbit one another, very fast, in quite an asymmetric way. One of the two fellows is a huge, angry red star. The other one is bright too, but much, much smaller. About the size of the Earth. And it is rather white. Do not be fooled, though. Despite the huge difference in size, the tiny one is the master here, not the red giant. The small white ball is what remains of the core of a star that exploded a few hundred million years before you arrived. As the star died, as it expelled its outer layers in all directions, its heart got compressed and became what now shines right in front of you. A *white dwarf*. An extraordinarily dense and hot object. Normal white dwarves take tens of millions of years to cool off and fade away, eventually becoming cold, dark, lonely space-wanderers. This one, however, has chosen a different path altogether.

To give you an idea of the density of a white dwarf, let's build a baseball out of different materials. A normal baseball, made out of rubber and leather and air, weighs around 5 ounces. The same volume filled with lead would constitute a ball weighing around 5 pounds. Filled with the densest element naturally occurring on Earth – it is called *osmium* – a baseball weighs about twice as much again: around 10 pounds.

Now, fill the same volume with material from a white dwarf and you get a baseball that weighs 200 tonnes. Within the kingdom of the extremely dense, white dwarves are ranked third. Just behind neutron stars (so called because they only contain neutrons) and black holes. One could hence expect extraordinary nuclear fusion reactions to take place within them, as within a star, but that is not the case. Unless they find a way to grow, that is. As a matter of fact, white dwarves remain white dwarves as long as they contain less than 140 per cent of the mass of our Sun.

But this one does have something to feed upon. A star. A red giant.

The red giant is being eaten alive, in front of your very eyes.

Gravitationally out-powered by the white dwarf's enormous density, the star is doomed. It can't even hold on to its own outer layers. As it orbits the dwarf, its surface is torn off to form a long trail of bright, burning-hot plasma that you can see spiralling down towards its greedy dance partner, creating a shining, twisted cosmic river meandering towards the white dwarf's surface, where it is harnessed and compressed.

Tremendous energies are at work here. Spacetime itself can feel it: gravitational waves are being created by the dance of the red giant and the white dwarf, creating ripples that propagate through the very fabric of the universe, altering space and time as they wash over nearby objects.*

And as you watch more and more of the giant star's matter fall towards the white dwarf's surface, you rightly feel that something extraordinary is about to happen. The white dwarf has indeed gained a lot of weight, reaching 140 per cent of the mass of the Sun, a mass-threshold. The pressure in its own core suddenly becomes

---

* If you wonder what effect a gravitational wave might have on you, here are some very roughly estimated numbers. Had you been near the two black holes whose collision was detected by LIGO, your size and width would have wiggled by about 2 or 3 per cent. It may not sound as being much. But so would have the size and width of everything around you. Including planets and stars. Now as they spread outward, these waves would have faded, just like waves created by a meteor hitting the ocean fade with the distance. As they reached Earth, 1.3 billion years later, the merging black holes' gravitational waves effect on spacetime was ridiculously small. We are talking about a variation no bigger than the width of a human hair over a length equal to the distance between the Earth and Proxima Centauri, the star you visited in Part One, the closest star after the Sun, four light years away. That is how good LIGO is. And such a wiggle is what was detected. Black hole collisions, of course, are extraordinarily violent events. The gravitational waves created by the white dwarf eating the giant star you are looking at now are much smaller. But such a meal is a violent event nonetheless. Especially towards the end.

sufficient to trigger a new, astoundingly violent chain-reaction that brings the white dwarf to an extraordinary demise. In the blink of an eye, it blows up. The explosion is 5 billion times brighter than the Sun. An impressive swansong.

Such explosions are called *type Ia supernovae*. They occur about once every century in any given galaxy. They are incredibly handy because they are all very similar. Identical, even: they always occur when a white dwarf reaches 140 per cent of the mass of the Sun, after feeding on another star, and hence they always shine with the same light: 5 billion Suns combined into one little spot not much larger than our Earth. Much brighter than the Cepheids, they are ideal candles with which to probe the furthest reaches of our universe, and to check Hubble's expansion law.

Type Ia supernovae are so much brighter than anything else that, unlike Cepheids, human-made telescopes can single them out in faraway galaxies. Knowing their intrinsic brightness, as with Cepheids, scientists can infer how far away they are, and how fast they are moving away from us.

In 1998, two independent teams studying such distant supernovae published their results. One was led by US astrophysicist Saul Perlmutter and the other by US astrophysicists Brian Schmidt and Adam Riess. Both teams found out that about 5 billion years ago, after more than 8 billion years of normal behaviour, the universe's expansion started to accelerate.

The scientific community was shocked.

And so should you be.

Not only was it not expected, only the opposite result seemed acceptable.

At large scales, it is Einstein's general relativity that rules everything, and Einstein's gravity, like Newton's, only allows for objects to attract one another. Whatever fills in the universe, be it matter or antimatter or dark matter, must hence, in the long run, slow down any expansion. Not accelerate it.

Perlmutter, Riess and Schmidt's observations, however, said otherwise and the only possible way out of this contradiction was for something very new to be introduced to account for such an acceleration. And this something had to fill the entire universe. And it needed to have an extraordinary property: it had to be acting as an *anti*-gravitational force, repelling matter and energy instead of attracting them.

For some unknown reason, this new force overcame all the other large-scale forces in our universe about 5 billion years ago. Before that, its effect was zero.

This rather puzzling energy has been called *dark energy* and, funnily enough, to account for its observed effects, there must be a lot of it.

According to modern estimates, a tremendous amount, in fact.

Three times more than the dark matter.

Fifteen times more than the ordinary matter we are made of.

As a result of their discovery that our universe's expansion accelerates rather than slows down, Perlmutter, Schmidt and Riess were awarded the 2011 Nobel Prize in Physics, and the whole energetic content of our universe had to be completely re-evaluated. Today, according to NASA's satellite estimates, it consists of the following:

Dark energy: 72 per cent.

Dark matter: 23 per cent.

The matter we know (including light): 4.6 per cent.*

Everything you've seen so far throughout your journeys corresponds to only 4.6 per cent of the total content of our universe.

The rest is unknown.

---

* The total does not add up to 100 per cent because there are always some uncertainties in the numbers obtained. Source: Wilkinson Microwave Anisotropy Probe (WMAP).

Unlike dark matter, however, the existence of some form of dark energy had been postulated in the past. About a hundred years ago. By Einstein himself. He had even called it his "biggest blunder," although today it seems that his blunder was to have called it a blunder.

You may remember, from Part Two, that Einstein did not like the idea of a changing, evolving universe. He much preferred to think that time and space were, had always been, and will always be as he experienced them. Unfortunately, his general theory of relativity, in its simplest form, said otherwise. It said that spacetime can, and does, change. To allow for the possibility of a universe that did not evolve, he found out that he could somehow change his equations by adding a new term, the only extra term they allowed. At the time, it was a bold gesture: Einstein's equations meant (and still mean) that the local energetic content of our universe is absolutely equivalent to its local geometry, so if one of these two can change, the other can too. Adding some new form of energy everywhere therefore changed the universe's shape and dynamic everywhere. By energy, Einstein meant everything that has a gravitational effect, which now includes matter, light, antimatter, dark matter and everything else that has a normal, decent, attractive gravitational behaviour.

But the term Einstein added could have either effect (attractive or repulsive), depending on its value. It corresponded, physically, to an energy that fills the entire universe. He called it the *cosmological constant*.

Thanks to it, the universe could be static and well-behaved, abiding by his philosophical opinions.

Relieved, Einstein could sleep again at night.

About ten years later, however, Hubble's work turned the expansion of the universe into an experimental fact. No more static universe. So Einstein withdrew his cosmological constant and called its introduction his biggest blunder.

About one hundred years later, it now appears, rather ironically, that what he erased on paper could be the much-needed tool theoreticians require to explain the largest mystery ever uncovered by mankind: the dark energy that drives the acceleration of our universe's expansion. The cosmological constant can make the universe the complete opposite of static, undergoing accelerated expansion. It could solve the dark-energy conundrum. The only problem would then be to figure out its own source. We'll get back to that in Part Seven.

In the meantime, I wish pretty much everybody could make blunders like Einstein's.

Whatever it is, the idea of dark energy has already changed our vision of cosmology. Before Perlmutter, Riess and Schmidt's discovery, our universe was thought to have two possible futures, depending on its overall contents. Too much matter and its expansion was doomed to reverse at some stage, gravity overtaking it all, as if too strong a spring were attached to everything now moving apart. Within such a scenario, the whole universe would then contract and everything would end in what was called a *Big Crunch*. It's like a Big Bang, but the other way round, with time fast-forwarded, not rewound.

The other possibility was that there wasn't enough matter or energy to keep everything from moving apart. Perlmutter, Riess and Schmidt's introduction of dark energy says that this is the most likely future. Unless another surprise hits our telescopes one day, the chances are that this antigravity force field will make sure the expansion goes on for ever, leading to very cold cosmic tomorrows. Both (Big Crunch and freezing death) are rather gloomy perspectives, I agree. But as you shall see in the next and last part, that cold death may not be the end at all.

Now, again, it is also possible that Einstein's theory simply does not apply at these enormous scales. In that case, we are not allowed to use his equations to infer the existence of dark energy. Just as Newton's ideas tried out next to a big star lead to erroneous orbits, Einstein's equations could very well drift away from reality at some stage. As of today, however, the chances are that dark energy is real, and there is even a possibility that it has a quantum origin. A very exciting prospect for all those who like to link the very small to the very big.

In any case, whatever they are, dark matter and dark energy are a big deal. Newton's gravity helped us to find new planets around the Sun. Einstein's gravity has led us to mysteries far bigger, mysteries so big they may hold clues about, or keys that open doors to, unknown realms of our large-scale reality.

With the necessary humility such discoveries impose upon us, it is now time for you to see why general relativity cannot be a theory of Everything, and why it predicts its own downfall.

6 | *Singularities*

Remember the quantum infinities?

Remember the catastrophic consequence on spacetime of an infinite number of particles appearing everywhere, all the time, in the vacuum of quantum field theory?

To deal with it, scientists had to turn gravity off and make do with those infinities as if they weren't there, or by ignoring what lay in the smaller still. Then it worked extraordinarily well, as long as gravity was not quantum.

Now, let's leave quantum stuff aside for a little while more.

What about gravity on its own? Is it possible for the matter we know, the classical matter we experience on a daily basis, to have the same impact on the fabric of our universe? Can it make space-time collapse in on itself?

The answer is a definite yes. And this time we even see the result in the sky.

The image of many very heavy marbles thrown on a thin rubber sheet works well here.

Due to the bending they create, close-by marbles should roll closer to one another, creating a lump that bends the rubber sheet still further. With each new marble rolling down to join the group, the rubber gets more and more distorted.

At some stage, either because all the marbles have fallen in or because the remaining ones are too far away, this should end.

Nothing strange about that.

But if the rubber sheet is soft like chewing gum, if it is not strong enough to hold the lump of marbles in equilibrium with its own tension, it may well keep bending and bending, even if no more marbles fall in, until it breaks.

No material is strong enough to hold *any* weight. Hence the idea of a density threshold: put too much weight upon too soft a surface and the surface around the mass will distort and distort and distort, and eventually break.

Now what about spacetime?

Although it is not supposed to break, spacetime reacts to very high densities in perhaps an even more dramatic way, because the fabric, in this case, is not rubber, but space and time themselves.

Spacetime. Not a flat cloth, a volume. Plus time.

Spacetime bends and curves and stretches around the objects it contains, be they matter or any other type of energy. That's Einstein's understanding of it.

Keep piling energy (whatever its form) into a given volume and, like the rubber sheet, you are bound to end up with a problem. Beyond a certain threshold, nothing will be able to stop the bending of spacetime from getting steeper and steeper and steeper, even if nothing more falls in.

As the bending gets worse, whatever started the bending gets squeezed, making the density in there even higher, a vicious circle that inexorably leads to a collapse of spacetime, a collapse marred with infinities that general relativity cannot handle. Such infinities are called *singularities*. They are not the same as the quantum infinities you saw earlier. They have nothing to do with quantum processes. They occur when there is too much mass, or energy, in too small a volume. They are localized. And the possibility of their existence announces the breakdown of Einstein's theory of gravity.

In the late 1960s and early 1970s, when pretty much everyone else was either high, listening to psychedelic music or looking for new fundamental particles, UK mathematical physicists Roger Penrose and Stephen Hawking proved, in a set of celebrated theorems, that such collapses necessarily happen within a universe ruled, on large scales, by general relativity. With their theorems, they showed that Einstein's general theory of relativity had the very humble characteristic of predicting its own downfall.

Just as Newton needed a larger theory to account for Mercury's drift, it became clear that Einstein's theory needed to be expanded, if only to account for these collapses.

Where do they occur? you wonder. Can they be found in nature or are they merely theoretical fancies?

They are real, and I know you know where to find them.

One such singularity, the mother of them all, lies in the past of our universe, when our whole universe's energy was confined in a dramatically small volume.

In a sense, our universe was born out of such a singularity, since it is out of it that space and time became what they are today.

Another singularity lies deep within all the black holes that stud our universe.

Contrary to what many may believe, black holes are the opposite of empty holes: they are born when, due to some catastrophic collapse, *too much* matter ends up being squeezed within too small a volume. As you will hear later, the death of a giant star can trigger such a process.

The question that has both tormented and excited many brilliant brains ever since the Penrose–Hawking theorems, therefore, is this: since singularities apparently happen in nature, how can one even *conceive* of what happens within them? How can one even think about places where space and time do not make sense any more? What theory can be used to probe those catastrophic collapses?

A theory that involves both the very big and the very small.

Since black holes and our universe's origin both consist of an enormous amount of matter and energy confined to a very small volume, the answer should involve a theory that mixes gravitational and quantum processes.

Whatever theory we can find to understand our universe that is better than Einstein's, it has to include quantum aspects of gravity, i.e. spacetime.

Penrose and Hawking proved that Einstein's theory of gravity has deep limits, that it cannot explain all of our universe, neither in the past nor as it is now: it breaks down before one can reach the birth of spacetime, and it breaks down before one can probe what lies at the bottom of today's black holes.

That being said, one could think that all the blame for the difficulty of finding a quantum theory of gravity should be on gravity, on Einstein's baby. But you saw that that is not the case. There are problems with the quantum vision of the world as well.

Still, however hard it may be, you are now about to try to mix them both, for it is time for you to probe a black hole.

Considering the situation, you feel extraordinarily normal.

You are not ethereal, you cannot see through yourself, and your arms and legs and everything else in your body responds positively when ordered to move. You are flesh and bones and blood, and your heart beats the way it usually does. A slight pain in your neck seals it: you feel just like you do on Earth. But you are in outer space. Your robot guide, complete with its tinny yellow casing and tube for delivering particles, is right next to you, as tangible and real as you are.

You look around.

The futuristic airport is gone. You recognize nothing, but you guess you must be within a galaxy, near its centre. Billions upon billions of stars are shining as they normally do. Everywhere. Except right in front of you, where a dark patch of spacetime is devoid of stars.

As you move alongside the robot, you realize that the area of darkness is drifting against the background stars.

So it is close.

A void hanging in space. A dark threat looming over one and all. You know what it is.

It is huge, about 10 billion times the mass of our Sun. But this black hole looks nothing like the one you saw at the centre of the Milky Way. There is no ring of burning lights surrounding it. There is no nearby star about to fall in. This black hole has already gobbled up and digested all the stars that were once nearby. And pretty much all the debris as well. It is clean now. It has nothing to feed upon but the occasional incoming rocks deflected by some faraway misfortune. Some of them are on their way now.

"If even a hint of quantum gravity lies down there, we are going to find it," announces the machine.

"Is it going to be dangerous?" you ask.

"Of course it is. It's a black hole."

You look again towards the black hole, comparing it to the one you met at the start of the book. There is no jet of light bursting from its poles. There is just a rather circular, flat-looking black patch of emptiness. You are spiralling down the spacetime slope it creates. As you fall, the images of faraway stars passing near to its edge look distorted, and nowhere near where they were a split second before. From being points of light, they become small strings of brightness covering the dark disc's outer edge. And then they vanish, as if swallowed by the dark emptiness, before reappearing on the other side – where the distortion sequence is played again, but backwards, until they look like distant shining dots once more.

Light, it seems, is distorted by this hole, a hole that apparently extends from within, like a dark well, while its rim acts as a distorting lens.

With the robot by your side, you continue to spiral down. You are still rather far away from whatever the black hole is, but you already feel a sense of doom and you suddenly wish that whatever the robot intends to show you arrives soon, so that you can leave before it's too late – whatever "too late" might mean.

"Look over your left shoulder," says the robot, after a moment of silence.

You turn around. A rock is aiming straight for the black hole. It is a spinning asteroid the size of a mountain. It shoots by with astonishing speed, about 60 miles away from you.

You lock your vision on its dark silver surface, the only object moving against the black disc of the black hole.

The apparent size of the rock shrinks as it moves away. It is now about the width of a peach held at arm's length. Now it is the size of a small, distorted nut and then suddenly, as your spiralling fall brings you to the other side of the black hole, two images of the rock appear. One to your left, the other to your right. The distortion of spacetime around the dark patch is such that light seems to be able to take several paths to reach your eye . . .

"The rock will soon fall through," says the machine, almost regretfully.

"Fall through?" you enquire, all the more worried. "What do you mean 'fall through'? Through what?"

"Through the horizon."

"The what?"

"The *black-hole horizon*. The limit of no-return. You will see. Or you won't. No human or machine has ever been so close to a black hole, let alone inside one. There is a theory about what *should* happen down there. But it may be wrong. Crossing the horizon, we will be beyond what is known."

"Maybe we shouldn't get too close then," you suggest.

"Or maybe we should," responds the robot. "That's research. We will have to take some considered risks."

"Where should I look for the horizon, then?"

"Everywhere."

Moving its throwing tube to the left and right, the robot alternately points towards two opposite places near the black hole's edge, towards the two images of the rock, and in between.

Your eyes now moving from one image to the other, you wait for them both to continue their fall, to disappear through the horizon, into the hole. But by the time you've completed another full orbit, the small, nut-sized, silver-brown asteroid is still floating above the dark emptiness. Strangely, it doesn't seem to have changed in the slightest since last time you were above it. In fact, it doesn't seem to be moving or spinning at all any more.

"It didn't fall!" you shout, relieved that perhaps you're not doomed to be torn apart by a black hole today after all.

"It did," corrects the robot. "It is not there any more."

"Very funny."

"It is gone," insists the robot. "Only its image remains. That is the distortion of spacetime in action. Of space *and* time. Our time, yours and mine, is not ticking like the rock's. The asteroid is beyond the horizon. Its image is still on the horizon. That's how it is."

As you take this in, another object races past you, into the void: a glittering stone this time. It almost looks like an enormous diamond – and indeed that is exactly what it is. For some stars, when they die, can leave behind diamonds the size of a moon.

As you watch it fall, you complete another circuit around the black hole, and you realize that you are much closer to it than you were before. And moving much faster. Turn after turn, the asteroid's several images, with the diamond's now next to them, all seemingly frozen above a surreal darkness, become more and more distorted. And so does the rest of what you can see.

Whatever your eyes might be telling you, the robot is right again: the asteroid and the diamond are both absolutely beyond retrieval. And the black hole grew in size as it swallowed them both. Or at least its horizon grew.

"Is that what you wanted me to see?" you ask the robot. "That an empty hole grows as it swallows stuff?"

"Black holes are not empty at all," replies the robot ominously.

In fact, black holes are the polar opposite of empty: they are what happens when there is too *much* matter and energy in too small a space. To create one, tremendous energy is needed. As far as we know, only the most immense of shining stars release sufficient energy, at their death, to compress their heart into one.

You encountered white dwarves earlier on in your journey, and white dwarves are the results of similar compressions – but they are not as extreme as black holes. All types of such remnants of stellar collapse are impressive, but black holes are beyond them all. And while we're at it, as you spiral down a couple more times around the black hole towards which you are inexorably falling, let me give you another reason why they are quite so frightening and mysterious.

Were you to sit on any object in the universe, be it a rock or a planet or a star, you'd be able to send some light away to signal your position. But the denser the object you were to be sitting on, the more energetic your signal would need to be for it to climb up the slope created by the object in the spacetime around it. It is just like with the salad bowl: the deeper it is, the faster you need to throw a marble from the bottom for it to roll all the way up, and out. Sitting on a planet, a star or a white dwarf, you need successively more and more energy for a signal of yours to escape their pull and reach outer space without falling back.

Black holes are even worse. They contain so much matter and energy, and therefore create a spacetime slope so steep, that anything clumsy enough to approach them too closely is doomed to fall in. According to general relativity, nothing, in our universe, then has enough power to escape the black hole's gravitational grip. Not even light. The point of no return beyond which nothing can ever come out – the black hole's *horizon* – lies where the images of the rock and the diamond appear to be frozen, when looked upon from way outside.

The darkness keeps growing in front of you, as if a huge mouth was ready to swallow your reality.

The distant stars, everywhere, now look very different. You even have the confusing sense that what you are seeing ahead of you

actually lies behind . . . Turning your head around, you realize that
it is not just a feeling, it is actually the case. The light emitted by
the stars shining behind you, travelling as fast as light always does,
overtakes you and shoots by the slope created by the black hole.
Whatever rays propagate to the left of the monster reappear to its
right after having made a rollercoaster-style U-turn behind. And
then these lights shoot towards you, and hit your eyes. Looking
forward, you are also seeing behind you . . .

From where you are, in fact, you can see the entire universe just
by staring ahead.

And as you keep spiralling down, things get more confusing still.

The images of the rock and the diamond are now moving again:
as you get closer to them, your time and theirs get closer, and
closer, and they suddenly vanish altogether.

You've just seen them cross the horizon, something they probably did hours ago, according to their own watch.

Next to you, the robot has turned around, its throwing tube now
pointing towards outer space.

You slowly turn around too, fearing what you might find.

And what you see is beyond the imagination.

All the stars, everywhere, which a second ago seemed to be so
still, are now moving. Their lack of stillness, normally unnoticeable
even over the span of a human life, is now apparent to you. From
the closest ones to the furthest, they all shoot through space and
time. Some of them are so fast-moving that they even leave a trail
on your retina, drawing evanescent curves of light across your
image of the universe. Just as, when you were travelling closer and
closer to light speed earlier, shooting through the universe, you
watched an astronaut's life, and her children's and her children's
children's, race past, their time accelerated compared to yours. Back
then, your time and theirs were different because of your speed.
This time, it is all due to gravitation, to the bending of spacetime

caused by the black hole's presence. For here, around the black hole, your time flows slower than it does everywhere else. You are watching the future of the universe as it unfolds, and this is again what is meant, in practice, by space and time being united in space-time.

"Have we crossed the horizon?" you suddenly ask, worried. "Are we doomed to fall in for ever?"

The robot turns around again, to face you, and you realize with great surprise that its throwing tube has widened. In fact, it looks as if it wasn't made to throw particles any more, but bowling balls . . .

"*We* haven't crossed the horizon yet, no," it replies. "But *you* are about to."

If you didn't know better, you'd say that you detected a hint of delight in the robot's voice. But before you can react, it fires a heavy ball straight at your chest. Unable to avoid it, you have no choice but to grab the projectile. Instantly its velocity pushes you downwards, towards the gaping darkness . . .

You shout, you frantically try to grab something to stop yourself from falling, but there's nothing around to hold on to.

You fall. The robot is moving away.

A second of yours already corresponds to a minute of its.

And now an hour.

And now a day.

And now a year.

As the robot recedes into the distance, millions of years pass before you. Stars explode. New stars are born. And you see it all.

Billions of years out there are now gone. Another galaxy merges with the one you are in.

The robot is nowhere to be seen any more. You are on your own.

And you panic.

You have crossed the black-hole horizon. Dumbstruck, you watch the future of everything. Gripped with fear, unable to focus, you fall feet-first, eyes locked above your head on the unfolding life of the whole universe, while you disappear into an abyss of unknown nothingness at the bottom of which lies a singularity.

And now you turn around to stare into it, into the black hole's mysterious heart, where the opposite of nothingness, the very matter that creates all this nonsense, should somewhere create this absurdity.

To your great surprise, you do not see anything at all. Not even your body. No feet. No nose. Not even your own hand.

Light may fall upon you from above, from outside, but nothing rises from down there, from any direction whatsoever, however close. Light hasn't enough energy to do so. You have crossed the black-hole horizon and you are now doomed to forever plummet towards the surface of many collapsed star cores reunited in an endless imploding fall, until they stretch spacetime too much for Einstein's general relativity, with consequences unknown.

In fact, were you to really be there, you'd be dead, for if even light cannot make the tiny journey from your feet to your eyes, there's no way your blood could climb up the spacetime slope you are sliding down to reach your brain.

But since we still have much to see, we'll assume you are still alive.

Reluctant to stare into this bottomless darkness, you decide to turn around again, to look at the universe as its images flow down towards you, through the now-distant horizon. But you can't. Any movement that involves having part of your body budge up, towards the "above," the outside, is prohibited. It would require an energy that even light does not possess.

No upward movement allowed.

As you begin to wonder if anything could be worse than this, tidal forces begin to make your body ache. The gravitational effect of the black hole's invisible presence is now starting to drag your feet down more than your arms and head. The black hole's gravity is stretching your body. You are going to end up strung out like spaghetti.

Even if the treacherous robot had equipped you with the most powerful rocket thrusters ever invented, it would not have changed a thing.

Whatever the engine, were you to try moving upwards from within a black-hole horizon, you'd feel like you were powering along on the slippery, stretching fabric of spacetime, as if exercising on an endless running machine whose speed always exceeds yours, by a huge margin, dragging you back.

According to Penrose and Hawking, you are being pulled by the spacetime singularity that lies down there somewhere, a singularity that will never be seen from outer space. No light being allowed to escape the horizon, the singularity is hidden by it. Down there, the very notions of space and time break down, just like at some time before the Big Bang. No one could ever look into the heart of a singularity and emerge to tell the tale. Such places, it seems, must remain forever cloaked.

According to general relativity, neither you nor any atom that belongs to you will ever get out of there.

A sad thought, especially now that you are completely torn apart, reduced to a long filament made of all the particles that constituted your body.

A sad thought, yes, but general relativity should not be trusted down there.

For we must remember that general relativity is not a theory of quantum fields.

And the moment this thought occurs to you, hope immediately returns to your mind and you turn yourself into your mini-you state.

And you wait.

At first, nothing happens.

And then, surprisingly, you see all the elementary particles you were made of disappear.

Or jump, to be more precise.

Quantum jump, in fact.

And now they are out.

Out of the black hole, where, rather fortunately, they reassemble into a mini-you.

And the robot is there to meet you.

At this point you are tempted to lunge for it and attempt to break off its metallic tube for having fired you through the black-hole horizon, but before you can act, the robot's metallic voice announces:

"I've been waiting for you for about 10 billion years. I'm glad you recognize me."

You suddenly don't have the heart to hurt it any more. And besides, there are more important things to think about. Not least the fact that what you have just experienced is an instance of gravity and quantum fields interacting with each other.

All around, the stars are again moving imperceptibly slowly. Ten billion years really have passed since you crossed (sorry, were pushed through) the black-hole horizon. You look at the black patch of space from which you miraculously escaped. At first glance it doesn't seem to have changed a bit; but now that you know what to look for, it is as if yet another veil has been removed and you really *see*. Particles are escaping the black hole, moving away from it, radiating, as if the dark monster were evaporating.

Maybe this was happening all along, you realize, but you just hadn't noticed it. But how can that be?

As Richard Feynman once said, one only really understands a phenomenon when one can give many different reasons for it to happen.

So, while you and the robot watch the particles shower out into space, I'll give you four reasons why black holes are leaking particles. They are all linked to a process you've already encountered.

The first one is the simplest.

Quantum particles can borrow energy from their field, as you know. And they can do that when inside the black-hole horizon too. With this borrowed energy, they are allowed to move faster than light for a little while. Not for long, but for long enough to quantum jump out of the black hole's no-return zone. That's what you did as your mini-you. This is a quantum process.

All the ways to understand what happened to you are quantum in essence, so they all come with the usual health warning, for like so much of what you have seen in the quantum world, they might sound absurd.

The second reason is no exception: you could say that all the particles that fell through the black-hole horizon also did not fall through it. Did, and did not. Out of all the possible paths a particle (understood as a wave) might take to fall in, most of them are a miss, because there is more space outside the black hole than in it. Amazingly, this idea, worked out carefully, makes the black hole evaporate in the exact same way as the first reason above.

A third reason goes as follows: due to the horizon that separates them from one another, the vacuum inside the black-hole horizon is different from the vacuum outside, so that some form of vacuum force, a Casimir effect, should push the horizon inwards, making the black hole shrink and evaporate. This again, rather miraculously, gives the same result as above.

The fourth and last reason I'll give here is that particle–antiparticle pair-creation occurs near all black-hole horizons, with antiparticles falling in more often than particles, just as more often than not there are more particles than antiparticles around us. Having crossed the horizon, the antiparticle is then bound to end up annihilating with a particle there, making them both disappear, while only a particle remains outside: the particle that was created with the antiparticle, the twin of the particle that got annihilated inside. Again, this gives the same result.

These are all quantum effects you've witnessed before, but they are here applied in the vicinity of a black hole. And they all lead to the same conclusion: black holes evaporate. They leak stuff.

As you now watch the black hole shine, therefore, you realize that this cosmic monster of a black hole, which had been swallowing entire stars for eons, is not black any more but grey. And shrinking.

Even more surprising is the fact that the more particles it shoots out, the hotter it seems to become, and the hotter it becomes, the more particles it shoots out. A vicious circle that should inexorably lead to its death.

The death of a black hole.

Unbelievable as that might sound, the black hole you are watching *is* shrinking, and emitting some radiation. The energy spacetime stored within it by swallowing whole worlds is now given back to outer space, a particle at a time, as if, like radioactive decay, black holes were around to break things down, to give the particles a second chance . . .

All the quantum fields of nature, excited by what is nothing less than the most powerful gravitational object known in the universe, are now using this unexpected bonanza to stuff themselves with energy. As the black hole gets hotter and hotter, their fundamental particles – particles that had until now lain dormant – wake up and shoot away. You see it happening. And the smaller the black hole becomes, the higher the fields' excitation, the more energetic the particles shot away. Gravitational energy is transformed, once again, into matter and light.

As all this unfolds before your very eyes, you realize that it is most contrary to Earthly principles: a mug of hot water, on Earth, doesn't heat up as it evaporates. It usually cools down. Were it not so, forgetting a hot coffee on a table would lead to disaster. The evening news would be filled with stories such as: "Yet another cup of coffee ignites a table, setting a whole building on fire. Remember to always dispose of your hot beverages in the appropriate bins."

Black holes are apparently different from cups of coffee. The more they evaporate, the more they shrink, the hotter they get. No one knows what happens at the end of this process. Do black holes disappear with a final bang? Does a weird, tiny remnant with peculiar properties stay behind? To figure out an answer, one would need to know what laws rule the singularity that hides deep within. Scientists have been searching for such laws since 1975.

It was in that year that Stephen Hawking discovered, on paper, that black holes evaporate.

At first, he did not believe his own calculations. Light appeared to be coming out of a place where no light was supposed to shine at all. So he did his calculations again. And again. Only to see that light and particles really did manage to find their way out of black holes. He published his finding in the journal *Nature* and instantly became famous all around the world, beyond scientific circles. Quantum effects made black holes evaporate. Whatever falls in is *not* doomed to stay inside for ever. It gets out, although not in a recognizable way. Being able to evaporate, black holes therefore do behave as if they have a temperature, a temperature that is today known as the *Hawking temperature*.

As you watch the black hole shine the last of its energy away, you realize that what you are looking at tells you that the very big and the very small *do* talk to each other, as of course they should. Black-hole radiation is the only proof so far that our theories might reflect nature in that respect. It is *the* hint that says a theory of quantum gravity might be possible after all. Any serious contender to such a status will have to predict the Hawking temperature and black-hole evaporation – all the way through to a black hole's death.

"Black holes can die," you say out loud in disbelief.

"Like everything else in this universe," says the robot.

But around the end of the 1970s, Hawking's discovery also led to a very strange and rather unnerving statement. With his temperature formula in hand, he tried to extract and decipher, out of the radiation he had found, some information about what made a black hole in the first place. To make things easier, he started with an already fully formed black hole and threw in different materials, to see how they each might be affected by the subsequent radiation. Amazingly, there was no difference. Nothing within the radiation emitted told him anything about what he had sent in, apart from their mass. It seemed, from what he could tell, that black holes purely and simply bleached out all the characteristics of what they had swallowed. Except the mass, that is. Whether a few humans, a bunch of books or a rock or a diamond were to fall through a black-hole horizon, if they happened to have the same initial mass, they would later be evaporated in the exact same way. For black holes, it seemed to Hawking, humans and books and stones all tasted the same. For all of us, this means that as far as black holes are concerned, only our mass has any meaning, which may strike some as being a bit reductive. For scientists, however, this was a philosophical catastrophe.

Until Hawking's work, black holes were supposed to gobble up everything that crossed their horizon, and grow, and this was not a problem. Anything that falls in isn't lost. It is just stored behind a horizon and difficult (impossible, really, but never mind) to retrieve from outside.

With black holes evaporating bleached information, we are confronted by a troubling realization: things start to vanish from reality. *Hawking radiation** being independent of what falls in, these dark monsters become memory black-outs for our universe. And once black holes have evaporated their past away, what they stored is not merely difficult or impossible to access, it just isn't anywhere any more. It is gone. Science was looking for a theory of Everything, a theory to explain it all in one single formula, and the first result reached through such an attempt dealt a hell of a blow to science as a whole. Science, forever unable to account for such lost pasts that occurred in black holes, was told to give up on hoping to one day describe and understand our universe's entire history. Hawking radiation was not a bell tolling the end of quantum physics or general relativity, but the end of physics as a means of learning where our universe came from. This problem has been dubbed the *black hole information paradox.*

Today, physicists are more familiar with the crude approximations Hawking used to reach his result. But forty years after his discovery, when Hawking asked me to work on it with him, the problem remained shrouded in mystery. But there are now hints that a way out may have been found, for if we apply what is known

---

* The Hawking radiation is the name given to what gets out of the black holes when they evaporate.

about the quantum world to black holes themselves, then black holes could be there, and not be there . . . Where such ideas have led scientists is what you will discover in the next and final part of this book.

For now, however, from some unknown billions of years in the future, you suddenly remember the robot's suspicious happiness at seeing you reappear out of the black hole. Didn't you wonder at the time why it was so pleased that you recognized it?

You thought it was genuine, didn't you? But it probably wasn't, and you now know the reason: the robot wasn't sure whether you'd remember anything at all. It didn't know if the black hole would or would not bleach your body and mind clean of all the information they contained. Then you recognized it, wanted to break it to pieces for having pushed you and it knew . . .

It knew that you remembered it all, that information was not lost in your case, even though you did not have the slightest recollection of flying backwards through the black-hole horizon.

You remember becoming a set of fundamental particles. And then being out.

In between, a quantum jump occurred, or something else.

Figuring out how precisely this could have happened is what a decent theory of quantum gravity is supposed to achieve. And since this is what you will soon start exploring, let me emphasize the fact that since the beginning of this Part of the book, you've entered a very theoretical world. Dark matter has never been created in the lab and nor has dark energy, and the same holds for black holes: their evaporation hasn't yet been detected by any experiment, directly or indirectly. Hawking would have received the Nobel Prize otherwise.

Black-hole evaporation, for one, is rather hard to detect.

How hard?

Let's see.

Take the Sun.

To turn it into a black hole, you'd need to squeeze it into a sphere 3.7 miles wide. That's equivalent to about two thirds the diameter of Boston.* Most black holes, throughout the universe, are born when giant stars die, so they should be bigger than that (the Sun is not a giant star). Now, let's assume that one of these "solar-mass" black holes has swallowed everything that surrounds it and now lingers quietly somewhere, far away from everything. Its radiating temperature, its Hawking temperature, should be about two 10-millionths of a degree above absolute zero (and absolute zero is about $-459.67°F$).

A 10-millionth of a degree is not much. It is hard to measure in its own right. But that is not the main problem. The main problem is that it is far less than the $36.91°F$ above absolute zero cosmic microwave background radiation that bathes everything in our visible universe. As a result, solar-mass black holes are not seen as evaporating now. In fact, to date they have never been seen to do so. They are and have always been masked by and feeding on the background leftover heat from the Big Bang era.

And since the heavier the black hole, the lower its temperature, it gets worse for the large, supermassive monsters that sit at the centre of most of our universe's galaxies. Their Hawking temperature is even cooler than solar-mass ones, not to mention that they are surrounded by extremely hot rings of infalling matter.

What would get Hawking a Nobel Prize might therefore lie in the world of the very small, since tiny black holes there should be very hot.

Unfortunately, we still have a problem: scientists are pretty certain they've spotted giant black holes, but they've never seen any

---

* In case you wonder, to turn not the Sun but our planet, Earth, into a black hole, you'd have to squeeze all its contents (you included) to the size of a cherry tomato.

tiny ones. Never mind, though. Let's assume they are there. Could we make anything out of them, in practice?

To figure this out, let me introduce a small parenthesis that will shed some light upon what I earlier called the Planck wall.

At the beginning of the twentieth century, one of the most impressive scientists of all time founded what we today call quantum physics. He was German, like Einstein, and his name was Max Planck. He received the Nobel Prize in Physics in 1918.

From his own discoveries, Planck understood that there was a scale beyond which quantum effects could not be neglected. Take a large object, and everything is fine. Newton's understanding of nature can be applied to it and whatever is expected of it corresponds to the reality we are used to in our daily lives. But shrink that object to smaller and smaller sizes and Newton's vision starts to fall apart. Newton, let me say once again, figured out a way to describe the world at a scale we are familiar with in everyday life. It agrees with our common sense. For the world of the very big and energetic, Einstein's vision has to take over. For the very small, it is Planck's. There, we must consider the quantum world. And there is a constant of nature that allows us to estimate when this occurs. It is called *Planck's constant*.

Planck's constant is on an equal footing with two other universal constants of nature, namely the speed of light, and the gravitational constant, which tells us how two masses attract each other.

One day, Planck started to play with these constants, and he built three things out of them. One was a mass. Another was a length. And yet another was a unit of time.

The mass turned out to be of 21 micrograms. Twenty-one millionths of a gram. It is called the *Planck mass*.

The length was a thousandth of a millionth of a billionth of a billionth of a billionth of a metre. It is called the *Planck length*.

The time was a millionth of a billionth of a billionth of a billionth of a billionth of a second. It is called the *Planck time*.

What do they correspond to?

They correspond to the scales beyond which neither gravity nor quantum physics can be used independently of one another. They are the thresholds beyond which quantum gravity is needed to explain what is going on, although some quantum-gravity effects may appear before these scales are reached.

What does that mean in practice?

Well, it means that the Planck scales give the size of the smallest black hole that can be.

So the smallest black hole today's science can imagine weighs about 21 micrograms. Funnily enough, it is a weight our minds can grasp. It doesn't sound that impressive. But it is huge when squeezed into the tiniest spacetime volume there is: a sphere that is a Planck length across. Such a black hole would evaporate away in a . . . millionth of a billionth of a billionth of a billionth of a billionth of a second. The Planck time.

Supposing we could measure such tiny things happening that fast, we'd need to create a Planck mass black hole to study it. But with our current technology, a particle accelerator powerful enough to create such a black hole by colliding high-velocity particles would have to be the size of our galaxy. Needless to say, that's way beyond our abilities and I doubt anyone is willing to start constructing such a device (apart from Hawking, for obvious reasons). Solace could come from outer space, though, where such tiny black holes might be detected as they fire away the last of their energy. But unless some as yet unknown phenomenon happens to tell us where to look, and what to look for, one would be extraordinarily lucky to spot one directly.

No one doubts Hawking radiation exists though. And that means a new reality looms down there somewhere: a quantum reality that contains space and time themselves.

And it is from this, as you shall now see, that the most extraordinary picture of our universe has emerged in the minds of some of the most brilliant scientists alive today.

Part Seven

A Step Beyond
What Is Known

1 | *Back to the Beginning*

As you've witnessed for yourself, the *visible* universe is not infinite and the Earth is, you are, at the centre of it. That is a practical fact, the key point being the word "visible": the light that reaches you from any one direction brings news from a past as distant as from any other direction, making your cosmic surroundings look spherical. It doesn't mean the whole universe is spherical, though, it means the portion of it that you can *see* is. The most ancient light that reaches you today left the surface of last scattering, the wall at the end of the *visible* universe, about 13.8 billion years ago, when the universe had cooled enough to become transparent. At the time of this last scattering, it is understood that the universe was about 380,000 years old, and 5,000°F hot. After that, it expanded and cooled down. Before that it was smaller, and hotter.

So, the visible universe is a sphere centred on the Earth, a sphere made out of all the pasts that reach us today. The outer edge of that cosmic epoch-layered onion, the edge of our observable pasts, is also the first visible part of it, the moment in our universe's history when light became free to travel unhindered by matter. You've been there. You've seen it. You've even crossed it. But there is something peculiar about it. Something very, very peculiar that you might not have noticed back then.

Do you remember that your billion-dollar-telescope-endowed friends found, when looking at the night sky, that the radiation that fills our universe is pretty much the same whatever parts of the deep night sky it originates from? This radiation, the cosmic microwave background radiation, trumpeted the triumph of Big Bang theory. It was the smoking gun needed to prove that our universe had been smaller in the past and much, much hotter. But neither your friends nor you paid attention to the fact that this radiation was far too uniform to fit what is expected of our universe's expansion. As you will now see, this extraordinary uniformity is one of the reasons why scientists introduced the idea of a cosmological inflation epoch that occurred before – and triggered – the Big Bang, 380,000 years before the universe became transparent.

And as you will now also see, this paves the way for the possibility of not one, but infinitely many Big Bangs.

Ask everybody in your neighbourhood to switch off their lights at night, and sit down on a deckchair to stare at the sky. Even though it is too faint for you to be aware of it, your eyes are receiving light from deep space, from the cosmic microwave background radiation. Looking for long enough, with the right equipment, you map that radiation, and you end up with a pretty uniform picture, showing a temperature of −422.76°F everywhere, 36.91°F degrees above absolute zero. Now, take your deckchair with you and travel to the exact opposite spot on the Earth. It is called its *antipode*. If you started somewhere in the US, you are now in the middle of the Indian Ocean. No lights around. You are on a raft, with your deckchair, staring at the sky again, gathering the light that pours in after it has travelled throughout the universe for 13.8 billion years.

Minus 422.76°F again.

The exact same temperature. The cosmic microwave background radiation.

But there's absolutely no reason for it to be the same everywhere. In fact, such a possibility should be ruled out . . .

The cosmic microwave background radiation that reached you in the US started from one side of the visible universe. That which reached you in the Indian Ocean came from the exact opposite direction. The sources of this light are so remote from one another (twice 13.8 billion light years apart) that unless something strange happened at some stage, there is no way, throughout the past history of our universe, that they could ever have been in contact.

So they shouldn't have the same temperature.

To realize how weird the fact that they do is, take a mug of hot coffee and bring it into your living room.

At first, unless you live in a furnace, your living room should be colder than your coffee, but if you wait long enough, the mug and the room will eventually have the same temperature. An equilibrium temperature, that is. As you've already noticed many times throughout this book, the coffee always ends up being too cold to taste good.

Now take your mug and put it in your fridge, with the door closed. A new equilibrium temperature will be reached after a while. An even colder one.

Travel to some hot desert with your beverage and yet another equilibrium will be reached. Warmer, this time.

All this should sound very normal. Nothing weird.

Now pour yourself another hot mug of coffee and put it back in your living room. It would be very unlikely for it to end up having the same temperature as the inside of a freezer in Japan.

Two objects or places that are not and never were in any contact whatsoever, objects or places that do not even know about the other's existence, have no reason to end up having the same temperature. That sounds like a fair supposition, doesn't it? So fair that it should apply to outer space as well.

For two opposite, antipodal parts of the night sky to have reached, after 13.8 billion years of separate existence, the exact same, −454.76°F temperature, they *must* have been in contact, somehow, at some stage in the past. But that is not possible: considering the age of the universe and its expansion rate, they are too far apart to ever have been in touch, by any means. Unless some very, very strange phenomenon occurred.

Something, for instance, would have had to travel faster than light.

Unfortunately, for a signal (meaning anything that could carry some information, whatever its form, from some place to another), that is impossible. We're not talking about quantum processes here, so signals, whatever they are, can't travel faster than light. That really is forbidden.

Still, the cosmic microwave background radiation temperature is what it is: far too similar everywhere for it to be coincidental. How can that be?

It could be that spacetime – the universe itself, that is – grew faster than light, at some stage in the past.

And that is what you saw when you travelled back in time beyond the Big Bang, when you entered the so-called *inflation era*, where the universe was filled with an inflaton field.

In its modern form, the idea of an inflationary early universe was first suggested in the 1980s by US theoretical physicist Alan Guth, Russian cosmologist Alexei Starobinsky and Russian-American theoretical physicist Andrei Linde. The basic idea is that a long time ago, even before matter and light and anything we know existed, beyond the visible universe, beyond the Big Bang, there was a field filling the universe with a repulsive antigravity force. That field was so extraordinarily powerful that it triggered a period of extreme expansion, an expansion that blew apart different parts of the early universe at a rate much, much faster than the speed of light, allowing for places that today seem too far apart to have ever been in touch to actually have been in the past.[*]

That is why the idea of an inflaton field was introduced.

---

[*] This represents no contradiction of Einstein's limit for the speed of light, by the way, because it is spacetime itself that expanded, not a signal that travelled that fast through it. Two objects moving away from each other at a speed greater than the speed of light will never ever be able to have any kind of conversation whatsoever.

But is it real? Can we, as with all the other quantum fields, detect some of its fundamental particles?

If it is real, most of its particles must have disappeared a long time ago (triggering the hot Big Bang), but it shouldn't have vanished altogether. Somehow, the inflaton field should still be around, filling the entire universe, lingering in one of its least energetic forms, a vacuum that, for lack of sufficient energy, hardly ever gets excited enough to produce and show us its particles.

Inflatons, as its particles are called, haven't been detected (yet). Still, many scientists are convinced that some sort of inflation scenario, with its inflaton field, must be rather close to what really happened, and since I personally like the idea very much, let's take it seriously and see what the history of a universe containing such a field should be like.

The inflaton field first did a very good job at separating different parts of our visible universe so fast that they have never been in contact ever since – and probably won't ever be again – although they were in the past.

Then the Big Bang occurred, with all its fields and particles and force carriers appearing out of the extraordinary amount of energy released by the decaying inflaton field, which subsequently became quiet.

Our universe's expansion then started. A normal expansion. Not a super-fast inflation.

The inflaton field did not vanish entirely, but too much of its energy had been used to trigger the Big Bang and it did not have any impact on anything any more until . . . 8 billion years later.

8 billion years after the Big Bang, after 8 billion years of our universe's steady growth, the matter the inflaton field had given birth to was diluted enough for its vacuum to wake up again, with a dramatic effect: its anti-gravitational power triggered an accelerated expansion of the universe.

The experimental detection of that acceleration, in 1998, is what got Perlmutter, Schmidt and Riess the 2011 Nobel Prize in Physics.

Of course, the way the inflaton field affects the behaviour of our universe now is nothing compared to how it blew everything apart before the Big Bang, during the *inflationary epoch*. Still, it may be responsible for what future awaits our reality.

Antipodal parts of the universe as seen from the Earth now are too far to ever have been in touch, but they were *before* the Big Bang. Antipodal parts of the night sky therefore have a reason to look alike as they do.

Now, is this introduction of a new field, the inflaton field, just a way out of a conundrum, a cunning trick to explain why antipodal points in the night sky have the same temperature, or did inflation really happen? Is it possible to check it?

Amazingly, it is possible.

2 | *Many Big Bangs*

Some time ago, you did an experiment with a cat. Schrödinger's cat. The idea behind it was to find a trick to turn some strange microscopic quantum behaviour into a macroscopic, observable reality. Well, inflation does that too. And there's no need for a cat here.

On a chronological scale, as you have just seen, the inflationary epoch happened before the Big Bang. The inflaton field turned what was an extraordinarily tiny universe into something macroscopic in an unimaginably small time.* The inflaton field and its fundamental particles (the inflatons) then decayed into pure energy, through $E = mc^2$. An extraordinary amount of energy was released and the universe became unbelievably hot. That is how the (hot) Big Bang is understood to have begun (within such a scenario), exciting the fields that were later to become those we and everything else are made of today.

During the inflationary epoch, the rate of expansion of the universe was so extraordinary that all the quantum fluctuations that could (and thus did) happen froze, one after the other. Even more extraordinarily, these frozen fluctuations can be seen today within the evermore precise picture scientists have of the cosmic microwave background radiation.

Inflation predicts the incredible smoothness of the background radiation that fills the universe. But that is one of the reasons why inflation was built in the first place. That is not really a prediction.

---

* If you are really keen on numbers, cosmological inflation is supposed to have taken place between around 0.00000000000000000000000000000000001 (10^{-36}) seconds and something like 0.0000000000000000000000000000001 (10^{-32}) seconds after the birth of space and time. During that time, the inflaton field made the whole universe grow by a factor of 100000000000000000000000000 (10^{26}).

But it also says that there should be some quantum fluctuations imprinted on this background radiation in the form of tiny temperature differences between one direction and another. Such differences are called *anisotropies*.

That was not a known fact, and yet such fluctuations *have* been detected: the American astrophysicists George F. Smoot and John C. Mather shared the 2006 Nobel Prize in Physics for experimentally detecting both the extraordinary uniformity of the background radiation *and* the minute anisotropies it contains.

These anisotropies are of the order of two thousandths of a degree Fahrenheit, but they are there all right. They are even thought to be what later triggered the formation of stars and galaxies.

Without them, the universe would be uniform. A star could never form.

Thanks to these fluctuations, there were tiny differences between one place and another in our young universe, and gravity then made those differences worse, amplifying them, creating the stars and all the other structures our cosmos is made of.

Now, inflation mixes the very small with the very big again, since it goes all the way from quantum fluctuations during the very early stage of our universe's development to the birth of the structures we see today in our universe. It even hints at what that mysterious dark energy might be, since this antigravity force could come from the leftover vacuum energy of the inflaton field.

Inflation potentially explains a lot of what is unexplained in outer space. So it has to be taken very seriously, and it is. And while we're at it, since I did mention a rather baffling consequence of such a scenario, here it is.

As it is understood today, the inflaton field cannot really stay quiet. It cannot be a "one-shot" field that happened, just once, at the birth of our universe. In fact, it is supposed to have triggered not just one Big Bang, but many. Infinitely many.

Like all quantum fields, the inflaton field should be subjected to quantum fluctuations, allowing it to jump locally from one vacuum state to another. Normally, for the fields you've seen so far, such a process leads to particles being able to jump from somewhere to somewhere else, or to appear out of nowhere. Here, however, it means that the field is able to create a small universe on its own. Or two. Or many. Everywhere. And when I say everywhere, I mean it, although the timescales involved may (or may not) be huge. This process is called *eternal inflation*. It never stops. Bubble universes appear within pre-existing universes, where the inflaton field's vacuum quantum jumped into another state, another vacuum. They are like drops of oil dropped on the surface of a lake. They grow. And grow. And grow . . . And other drops grow within these drops.

Bubble universes within bubble universes within bubble universes.

An example of a multiverse, but a multiverse of a different type from the ones you've already seen.* Within such a scenario, you and I would live in one such bubble universe and there may well be bubbles ready to appear within our spacetime at some stage in the far future, just as ours might have popped out of yet another bubble, one that is now much bigger and perhaps a bit damaged, or emptied. Our visible universe's potentially cold death in the future may hence be the mould needed for the growth of new bubble universes . . .

All right.

We'll have another look at these funny pop-up universes when you travel through the landscape of string theory at the end of this

---

* The first one consisted of all the parts of our universe that are beyond our observable reality, and the second was Everett's "many-worlds" interpretation of quantum mechanics. This is the third: universes born within universes.

book. In the meantime, eternal inflation may (and should) sound completely crazy to you (it does to me, but I like it), and yet, compared to the strings you are about to encounter, well, nothing will ever sound sane ever again, really . . . You should even consider the bubble universes you just encountered as an introduction to what is to be your final journey. Before getting there, though, before getting back to the visible universe and seeing where those famous strings might be hidden, what they are and what they imply for our reality, let's try and see if we can look beyond inflation using what we have learnt so far.

To those who ask "How did the universe begin?," the eternal-inflation scenario may not sound very satisfactory, for there isn't any beginning, really. It's bubbles, all the way.

But there may be other possibilities.

I cannot list them all for you here. I'll just mention one.

The historical first.

3 | *A Universe Without Boundary*

The inflationary epoch took place before the Big Bang.

With *eternal* inflation, infinitely many universes have been, are being and will be born, since forever, ours just happening to be ours. Now, let's just imagine *one* universe, with *one* "beginning" (whatever that means), with *one* inflationary epoch.

And let's rewind time, starting from the Big Bang.

There is the Big Bang: *Boom.*

And before was inflation. Looking at it backwards, it is a dramatic collapse.

And then, well, we hit the problem.

The Planck wall, the Planck era, when and where space and time stop making sense.

That Planck wall lies some 380,000 years before the surface of last scattering, the surface at the end of the universe, and, if we were allowed to make such a guess, about a Planck time after what we could call *time zero.*[*] But we are not allowed to make that guess. We cannot reach time zero from within our universe. We can't talk about a time where (or when) time did not exist. Talking about "beyond" or "before" the Planck era makes no sense. Quantum gravity is indeed needed for this, with its unknown load of new concepts to replace space and time with quantum somethings. A hard task, akin to finding an initial condition to the existence of our reality. Hard, but not impossible. Stephen Hawking and US theoretical physicist James Hartle tackled exactly that problem about thirty years ago. They were the first to do so. And here is what they did.

---

* If you do not remember and care to hear it again, the Planck time isn't much: a millionth of a billionth of a billionth of a billionth of a billionth of a second.

Picture your mini-self in a very young universe. A universe in which space and time have only just begun to make sense. It is tiny. A bit bigger than the Planck size, but not much. You are inside, you are tiny too.

And you cannot see much.

Anything that happens on a scale smaller than Planck's length is beyond space and time and is therefore hidden from your sight.

You are there, tinier than tiny, within an extraordinarily young universe, and as good as blind . . . but wait . . . doesn't this remind you of situations you've encountered before?

When visiting the quantum world, didn't you switch to yogi mode, eyes closed, in order not to interact with anything and access what was hidden from sight? When probing the inner parts of atoms, to guess what was going on around you, you actually *had* to somehow be in yogi mode. And to make sense of what you discovered that way, you learnt that in the quantum world, when nature and its cats are left unchecked, all the quantum possibilities happen simultaneously.

Here, it is worse.

It is not a cat or a particle that is invisible, it is our entire universe's past, a past that is hidden by a wall that marks the very birth of space and time as we know them. This wall, this Planck wall, is now everywhere around you and what lies beyond is inaccessible to your senses.

According to quantum law, the Planck wall therefore hides a superposition of all quantum possibilities.

The possibilities of what? you may wonder.

Well, of pasts.

It is the young universe itself, as a whole, that is hidden from view by Planck's wall, so it is the young universe itself that should there abide by one of the golden rules of the quantum world: as long as no one is looking, all possibilities can – and do – occur.

Hawking applied this idea to the very early universe.

But he could not use the time we know and use every day. No one is allowed to use it beyond the Planck scale. So he turned it into something else, easier to manipulate, called *imaginary time*. Using it, he then considered all the possible past histories of the universe, all the past histories that one cannot see from within.

The idea came to him in the 1980s.

He had just figured out ways to deal with quantum black holes. He knew they were grey, that they emitted particles. He knew quantum gravity had to exist. Hawking's mind was now staring beyond the Big Bang.

With his colleague, US theoretical physicist James Hartle from the University of California, Santa Barbara, he wrote down a formula that, for me, for ever changed the universe as it is apprehended by the human mind.

Hawking and Hartle assumed that all the universes that led to our present one must have appeared out of nothing (really nothing, a mathematical nothing), some finite *imaginary* time ago.

And they considered all the universes that had that property.

And they looked at them.

And there were many.

And they imposed the golden rule of the quantum world on them: instead of picking one out to subsequently evolve it into our reality, they took them all into account. On paper, this means they added them all up, with a plus sign, and they stated that the sum was what the universe we're in looked like "before" the Planck wall, where one could not look at it. Their mathematical formula is today known as the *Hartle–Hawking wave-function of the universe* and the initial condition, the one that says that all the possible universes to be taken into account are the ones that came to be out of nothing, is called the *no-boundary proposal*.

The universe, our universe, from their point of view, with all its possible states as a young universe, had no beginning.

And then it became ours, some finite imaginary time later, when space and time began to make sense.

What this exactly means doesn't really matter here.

The crazy thing is that they did it.

They wrote down a mathematical initial condition for the whole universe. They mathematically tackled the problem of our universe's creation out of nothing.

Now for some words of caution: this is not the end of the story. Keeping track of almost any calculation within the mathematical framework Hartle and Hawking came up with is, unfortunately, painfully hard (not to say impossible).

Still, just by writing it down, they became the first people to give a mathematical formula for the origin and subsequent evolution of our reality.

An extraordinary milestone for humanity.

Mankind has been trying to unravel the laws of nature for thousands of years.

Our comprehension of these laws has changed, and improved, ever since.

A hundred years ago, Einstein came up with a new vision about gravity and the rest of us began to understand that the past was to be found not only below our feet, by archaeologically digging into the Earth, but also among the stars. At about the same time, many scientists started to discover the strange quantum laws that rule the world of the very small.

And then, about thirty years ago, buoyed by the black-hole evaporation result, Hartle and Hawking boldly began to put it all together and develop a mathematical formula for the origin of everything.

Their insight may, in the future, turn out to be deeply flawed, of course, and the same can be said about all the ideas that take us beyond experiment, but it doesn't really matter. What matters is that the question of the origin of our universe has entered a new era, an era where mathematical physics is at least allowed to enquire about the subject.

Hawking's idea of considering all possible universes using a different (imaginary) time did not come out of nowhere, though. It is rooted in the works of some of the brightest minds of the twentieth century, namely Paul Dirac and Richard Feynman, who created such a concept to build our modern theories of quantum fields.

The visible universe, within such a scenario, is still a sphere roughly 13.8 billion light years in radius. It is the largest size we can probe. Still, it is funny to think, once again, that as one collects the light and signals pouring down on us from outer space, as one travels further and further in the very big, one ends up not only looking at the past, but also at the very small.

Our ancestors did not know that.

And as you will now see, the opposite may well also be true.

You are now about to travel to the very small again, but this time you will go further than you've ever been before. Down there, you will find a window opening onto a new reality altogether, a reality bigger than anything you've dreamt of so far. Bigger even than the bubbles within bubbles within bubbles of the eternal inflation.

In the big, you found the small.

In the small, you'll now find the huge.

But where should you look?

As you now know, our whole visible universe is a sphere 13.8 billion light years in radius. From such a gigantic perspective, one first sees filaments of giant clusters of galaxies bathed within gases and dark matter and, more fundamentally, all the quantum fields there are. These can't be seen from that far up, but they can be felt. They are the matter that makes up the visible universe. They are the Higgs field that gives mass to everything that has a mass. They are the inflaton field, or the dark energy that counters gravity's action and keeps the universe expanding quicker and quicker.

And there is also gravity itself, bringing everything closer to everything else.

You are out there, watching it all, and you start zooming in.

You see galaxies now, with their hundreds of billions of stars. Their supermassive central black holes spit jets of the most energetic light and matter there is. You see the dark matter's presence. You see it preventing the galaxies from being torn apart because of their own spinning.

Keep zooming in.

You are at the scale of stars, huge balls of burning-hot plasma shining the light that we humans use to probe the distant universe.

Then come the planets, spherical worlds too small ever to become stars.

Smaller yet, there are the asteroids, the comets, the living beings that our planet harbours beneath 62 miles of atmosphere.

And then come the microbes, the cells, the molecules, the atoms, the electrons and photons, the protons and neutrons, the quarks and the gluons.

Keep zooming in.

You are back into quantum-field territory.

Gravity, here, is outgunned by all the quantum forces.

You keep zooming in. And then you stop.

Do you remember what went wrong with quantum fields? Do you remember about renormalization, the trick quantum-theoretical physicists use to get rid of the infinities that plague their work? And do you remember that attempts to look at gravity as one would look at a quantum field utterly failed, for the infinities that occurred in that case could not be removed, by any means, making space-time collapse everywhere? It is these infinities that we shall now get rid of. Behind them, you will find the window that leads to the immense new reality I mentioned at the end of the last chapter. You'll cross that window very soon now. But we first need to remove those annoying infinities.

How are we going to do that? Well, let's see. What do we know about spacetime? We know that its description using early twenty-first-century physics has its limits. In the very big, that limit is some place beyond the Big Bang, beyond the inflationary epoch, when the universe was in the Planck era. This limit lies 13.8 billion light years away in space and time.

In the very small, the same limit exists. And it occurs everywhere.

Zoom in on anything and you should, at some stage, reach the Planck scale.

Unless something prevents you from doing so, that is.

We know, thanks to Hawking's work on black holes, that gravity is not shielded against quantum effects, that quantum gravity somehow exists, although we do not necessarily understand what that might imply for reality within its territory.

There is a limit to what we can probe both in the very small and in the very big, and that limit is given by Planck's scales.

Has any experiment reached those limiting sizes or energies or times in a laboratory?

No. None has. They are far too small, far too energetic, far too fast. As of today, it is a theoretical limit. And to make things worse, it is also a practical one, for one cannot in fact reach it.

Why?

Because a tiny Planck-sized black hole would appear in the process, the Planck-sized black hole I mentioned at the end of the last part. To probe reality beyond that black hole, one would have no other choice but to try to send in more energy, more light with shorter and shorter wavelengths, hoping it would bounce off something and reveal its existence to our eyes, but it would not. The light would be swallowed by the black hole, only making the black hole bigger, hiding the quantum gravity scale even more. In other words, as far as modern knowledge is concerned, what lies beyond the Planck scale cannot be probed.

So what do we do?

Well, we can again try to be smart.

And we can for instance suggest that nothing prevents quantum gravity, or some new physics, from kicking in *before* the Planck scale.

With the best modern particle accelerators, with the best use of what can be observed in the sky, theoretical physicists are confident they understand how nature behaves almost all the way from huge, galactic scales down to the scale at which all the quantum fields merge into one. The grand-unification scale. The energy needed for it is about 1 per cent of the Planck energy. It is huge, obviously. It corresponds to a temperature of around 100 billion billion billion degrees. But it is *not* the Planck limit.

Now, you probably remember that energy and size are related: the higher the energy of a wave, the shorter the distance between two consecutive crests. So, a hundredth of the Planck energy (1 per cent of it) does correspond to a size in the realms of the very small. A size that is 100 times bigger than the Planck length.

This means that there is an unspoiled territory of reality that stretches between at least 100 Planck lengths and the Planck length itself.*

Experimentally, nothing is known about what happens there.

A nice way to imagine what it feels like, for a theoretical physicist, to have this experimental gap is to think about what the world would look like if your eyes only allowed you to see it with a 1-yard resolution. Normally, you can see the world at such high resolution that you can spot objects far thinner than a human hair, but imagine not being able to detect anything smaller than a yard across. Probing your environment, you wouldn't see any detail anywhere. You wouldn't even be able to see babies. Children would suddenly appear, once they became a yard tall . . .

I'm not saying there might be babies smaller than 100 times the Planck length, but we do not know what nature might be hiding there. And our reality *is* rooted somewhere in the very small. That's what it is made of. That is what *we* are made of. And since no experiment has ever probed these scales, it is very possible that space and time start differing from what we are used to somewhere *before* the Planck scale. It is also possible that, because of that, the nature of gravity and matter and light start to change there too. Drastically, even.

It is, for instance, possible that they all become one.

Until now, you've seen what was mostly known.

Then you saw what problems arose from what is known.

You are now about to go way beyond.

---

* In June 2015, the energy reached by the Large Hadron Collider particle accelerator near Geneva smashed all previous records and almost halved that unknown. But we will have to wait for a year or two to hear about potential breakthroughs.

And we shall assume that it is all real, so that you can travel through it, but do keep in mind that this is pure theory.

Still, some of the brightest individuals of our time have worked for decades to bring you this picture.

5 | *A Theory of Strings*

A curious haze of blue electricity surrounds the silhouette of your robot companion, as if an inner excitation were spreading from its electronic circuits outwards. You are both floating in outer space, surrounded by faraway galaxies, near to the point where the black hole you escaped from vanished entirely.

You have seen everything there was to see.

You've flown on a very fast plane.

You've seen the vacuum fluctuation of quantum fields and got acquainted with matter and light.

You've seen stars exploding to create new worlds, white dwarves and black holes, which you've in turn seen evaporating, hinting at the existence of an as yet unknown theory of quantum gravity.

"Now it is time to probe even further," says the robot.

And at once you both begin to shrink.

You see particles fly by. Light shoots past. You see the vacuum fluctuations of all the known fields. And you keep shrinking. You are at the grand-unification scale, where all three quantum fields are believed to behave as one. You keep shrinking. You are far beyond your mini-you size. You'd need to blow up what is around you a billion billion billion times to end up the width of a human hair. Down there, at first, you see nothing. But then you do.

In front of you, there is something. A string. A string made out of nothing. Not even space or time. As you look at it, you even have the feeling that this very object you see wiggling replaces both these notions.

You haven't yet reached the Planck scale, and you won't be able to. In the theoretical world you are now entering, the Planck scale does not exist as you might have thought it did. But that doesn't mean that what you've seen up to now was wrong. It means that, down here, none of the concepts you've used can be trusted. Except quantum ones. But applied to strings, rather than to particles.

What is wiggling right in front of you now could be one of the most fundamental elements of the universe. It is a *quantum string*.

From its existence, it may be possible to explain everything you've seen before, including gravity. Including our whole universe's existence.

The quantum string in front of you is vibrating. Quantumly. You can't really pinpoint its edges, but you can tell they exist, although everything about that string is moving very, very fast.

It is beautiful, vibrating with a merry energy, and you feel attracted to it. Unable to stop yourself, you reach forward and, although it seems to be wiggling on its own anyway, you pluck it as if it were a guitar string.

Although the string is made out of nothing, you see many vibrations piling up, like harmonics on a musical instrument. The largest standing wave, on a real guitar, gives the main note. The others give the higher harmonics. As you look at the string here, it is like the blur of a guitar string . . . but without the guitar string itself. A string made of nothing, a fundamental string, if you will, able to wiggle. Remember that when the word "quantum" precedes a term from the vernacular, it is a clue that nothing is as it seems. Here, a "quantum string" isn't a string at all. The first vibration here does not give birth to a note, but to light. A particle of light. The carrier of the electromagnetic force.

All the quantum particles you've encountered before, all the particles that make up your body and all the matter in the universe, could be vibrations of such open strings . . .

Something attracts your attention to your right. You turn your tinier-than-tiny head to see another string, a different one. It is not like a guitar string, but more like a closed loop. It too vibrates. Quantumly, again. And its first excitation does not correspond to light any more but to a graviton. A gravitational force carrier. It is gravity, quantized. This loop, this closed string, by itself, tells you that you are travelling within a quantum theory of gravity. Put such a closed string anywhere you can think of and its vibrations will have the exact same effect as gravity. And you see no infinity lingering anywhere. The infinities that plagued quantum gravity are gone. For good. Because you got rid of the notion of where things happen in space and time. With dotlike particles in a smooth spacetime, it is easy to conceive of a specific place where they could collide. And quantum field theory, despite its inherent weirdness, also says that when particles interact, they do so at a specific position in space and time. With strings, that is not the case any more. With strings, particles *are* string vibrations. String vibrations *are* particles. Throughout their length and time. They are spread. When they interact, it is neither anywhere precisely nor at any particular time. It is all along the whole string. There is no "infinitely" small any more. And this is what removes all the infinities you encountered before.

This loop, this closed string, has gravity in it, so it is gravity. And you have light emanating from the open strings. Taken together, they therefore become a theory that unifies gravity and electromagnetism . . . Quantum strings are thus more than just a theory of quantum gravity. A theory of quantum gravity "merely" deals with gravity, in a quantum way. It doesn't care about the other quantum fields. The strings you are looking at here do.

So what about the other fields?

Could these strings be a theory of Everything, a theory that unifies gravity and *all* the quantum fields we know?

For this, they'd also have to account for matter.

Where is the matter? You can't see any. So why are these strings so special? Where is the weirdness in their existence? Why are theoreticians so excited about them?

You are right to wonder, and though with those two strings you've seen, the closed and the open one, you can already say a lot, a lot is not everything.

"Let's move on," announces the robot, and you both start shrinking down still further.

The open string is now huge compared to you. As you watch it closely, you start to see that there was more to it than first met the eye. What you are about to do, no human made out of matter will ever be able to do. But right now, you can. Remember this, though: to move beyond what is known, something must always be given up. And what you are going to have to give up here is the specialness of your universe, a universe you might have thought was unique. But not just.

To get from Newton to Einstein, you had to give up the idea that the universe was static, that it had always been the same, that gravity was a force. You had to introduce spacetime, with its three dimensions of space and one of time, the four of them being intertwined into a single entity that deforms itself around matter and energy. To get from Newton to quantum physics, you had to give up the idea of particles being dotlike. You had to introduce waves and fields and uncertainty and different histories. Now to get from gravity and quantum field theories to strings, you have to turn everything that is fundamental into a theory of closed and open strings.

But that would be easy. What you have to also give up here is the idea that reality is made out of four dimensions only. Strings cannot live in a four-dimensional spacetime. They require more room. They live in a ten-dimensional universe.

As you approach the string with the robot, you begin to see that above each and every point of what you thought was contained within our universe, there are six new dimensions of space making up a world of their own. It is from these little extra dimensions that all the matter we are made of is supposed to come.

If you are struggling to visualize four dimensions, let alone ten, don't worry. All you need to know is that the six extra ones extend in different directions from the usual left–right, up–down, front–back of our three-dimensional world, and are too tiny for you to feel their existence, or to travel through them, in real life. But the robot and you have now shrunk so much that you can.

What do they look like?

Well, it is impossible to tell. There are so many of them! So many possible ways to interweave extra dimensions and get a string . . . So many ways to wrap these extra dimensions around themselves, each different wrapping giving a different ground for reality . . . Theoretical physicists have even guessed how many possibilities there could be, and the number they reached is about 100,000,000,0 00,000,000,000,000,000,000,000, 000,000,000,000,000,000,000,000,0 00,000,000,000,000,000,000,000,000,000,000,000,000,000,000,0 00,000,000,000,000,000,000,000,000,000,000,000,000,000,000,0 00,000,000,000,000,000,000,000,000,000,000,000,000,000,000,0 00,000,000,000,000,000,000,000,000,000,000,000,000,000,000,0 00,000,000,000,000,000,000,000,000,000,000,000,000,000,000,0 00,000,000,000,000,000,000,000,000,000,000,000,000,000,000,0 00,000,000,000,000,000,000,000,000,000,000,000,000,000,000,0 00,000,000,000,000,000,000,000,000,000,000,000,000,000,000,0 00,000,000,000,000,000,000,000,000,000,000,000,000,000,000,0 00,000,000,000. All of them potentially able to give rise to a universe, although not necessarily a universe like ours.

A huge number of possibilities. A "1" followed by 500 zeros. The universe you and I were born into may just be one of them. Or many could be like ours. No one knows yet. It may well even be that these possibilities happen to *all* exist at some stage, within bubbles created by the eternal inflation you've just heard of, but that only a handful of them can create a universe where the laws of nature are compatible with life as we know it. For you to be, for you to exist, as a human, a particular set of extra-dimensional shapes must have been selected, or the laws of nature wouldn't allow our existence. How did that selection occur? No one knows either, except that they *have* to have been selected, in order for you to exist here, in our universe. Such a selection argument is called the *anthropic principle*. It states that out of the unfathomably many possible forms extra dimensions could take, only the ones compatible with the existence of humans need be taken into account for us to be here, or we wouldn't be around to talk about them. It is a nice idea. And it gets better. Instead of them all being tiny, one or more of these extra dimensions can be huge.

"Come with me," says the robot, beckoning with its particle-delivery tube for you to follow. "We may never see this again."

And the most extraordinary thing happens.

Since forever, you have been taught that it is impossible to look at the universe from outside of it. That talking about its edge, its border, is nonsense. The universe being all there is, by definition, it was meaningless to try and even picture what it might look like from above, or below. And yet, moving along a direction that is neither up nor down, nor left nor right, nor forward nor backward, the robot is now taking you out of it. Its edges, it now seems, do exist. But they are not within the dimensions your usual senses can perceive.

You are out of it.

You see it all.

Your entire universe.

From another dimension. And you see that the open strings, the shoelace ones, whose vibrations give rise to light, are now vibrating in many different ways, depending on the hidden dimensions into which they extend. And you also see that all these open strings' ends are stuck to your universe, the universe you just left, whereas the closed ones, the loops, the ones that vibrate like gravity, these strings are free to roam outside, to leave the universe . . .

And as you become aware of something behind you, you turn around, and you gasp.

There is another universe.

Parallel to yours, to ours. And you see closed strings move from one to the other, showing that they can communicate through gravity. These are the fourth type of parallel universes, the most impressive of them all. Such things are called *branes*, like membranes but without the "mem," to show that they can be more than sheets, more than two-dimensional. What you are seeing is one such brane, one other universe, but there may be many. And they can be of many different dimensions too. And they can all turn into one another and behave like the strings themselves, when the mathematical physicists that study them change the way they all interact. They can either be separate entities or they can all be different aspects of the same reality, a reality looked upon from different points of view. And all this may be one aspect of a bigger reality still, whatever "reality" may mean in that case. And some scientists, led by the brilliant Argentinean theoretical physicist Juan Maldacena, even showed that all this could be understood without gravity, as if every single universe here could be described by what happens on a boundary somewhere . . .

The truth sinks in. You are outside the universe.

And there are other universes around, everywhere, of different dimensions. And there are tiny dimensions upon which strings wrap themselves, within and around those universes, making them vibrate into the matter and light that is forbidden to leave their brane, their universe, your universe. Their ends are free to move within the dimensions you were born into, but they are not allowed to leave them.

From where you are, as you see closed string loops move from one brane to another, you realize that some energy might be able to leave your universe. You even see what you believe might be black holes linking nearby branes through a tube of distorted space-time, with gravity from each brane attracting the others, and you suddenly wonder if, by any chance, there might be other people living in those other branes . . . Could black holes be a passage between your world and theirs? Could the singularity you did not reach open up on another reality? Could the birth of our brane, of our spacetime, be linked to collisions with other branes that existed before? Could the dark matter, the dark energy, be explained by the existence of branes?

Turning your gaze back to the universe you just left, it suddenly appears that something has happened to the flow of time, and you see bubbles of new inflationary universes pop up everywhere within yours, within your brane, spreading within what was your world like drops of oil on the surface of a pond.

"We should go back!" you shout.

But you are alone.

The robot is nowhere to be seen any more.

And you slip inside the brane nearby, hoping it is the one you came from.

And you start to grow.

The other branes are invisible again, and the strings that may make up your reality vanish in the distance.

The quarks and gluons are now around you. Now the protons and then the electrons, the atoms. The molecules. Dust. Sand. The sea.

You open your eyes.

You are on your deserted beach.

At the exact same spot you started your journey from.

The stars are shining.

A gentle breeze blows scents of exotic flowers your way.

Your friends are around.

They smile.

"He's awake!" says one. "Pour him a drink!"

You sit up, confused.

The drink arrives.

You pinch yourself. It hurts.

You take a sip.

You stare at the sea, the trees, the stars.

Shapes.

Shapes are appearing up there in the night sky. Faces.

Newton. Maxwell. Einstein. Planck. Schrödinger. Dirac. Feynman. Hawking. 't Hooft. Weinberg. Maldacena. Witten.

And countless others.

All smiling. All looking at you.

You want to talk to them, but instead they turn around, to stare at the majesty of outer space.

And then they all vanish into the stars.

And the stars themselves vanish, and so does the sea.

You blink.

You are back home, on your sofa.

Your window is open.

You sit up. You look around.

Your coffee is still there, on the table.

You pinch yourself again. It still hurts.

You take a sip to wake your mind up.

The coffee and your living room have reached an equilibrium temperature.

You spit the cold coffee out.

"I'm . . . I'm all right," you say out loud, but stretch to reach your phone and call your great-auntie, just to make sure.

And then you blink again.

Epilogue

Throughout history, philosophers—and now theoretical physicists—have been trying to picture the world in their minds. To unravel its laws, the laws of nature, laws whose existence is apparent to us all (but whose language remained hidden from us for a very long time), they projected themselves into situations that were not possible physically or experimentally. Such experiences are called *gedanken experiments*. They are experiments of pure thought.

It is a succession of such *gedanken* experiments that you have experienced throughout this book. They allowed you to travel, in thought alone, through the universe as it is known today, and beyond.

Schrödinger used such a process to show how strange quantum rules should appear when linked to macroscopic, everyday events. He ended up with a cat neither dead nor alive, but dead *and* alive. Weird stuff indeed, but now proved to be correct.

Einstein also made much use of *gedanken* experiments. He imagined what reality would look like were the speed of light to be a fixed speed limit. To do that, he sat on a photon. Looking at the world from there, in his mind, he came up with his theory of special relativity, which notably tells us that a plane travelling as fast as the one you were in really would end up 400 years in the future. This has also been proven to be correct. And he went on to tell us what gravity is all about, leading to discoveries of mind-bending proportions, even a century later. Intuition, although not based on the common sense that has allowed our species to survive so far, is what has driven research for more than a century now.

On February 11, 2016, a scientific paper signed by more than 1,000 scientists from all around the world announced that humanity's ability to peer into our universe's past and present has entered a new era.

For the first time ever, waves propagating through the fabric of our universe have been detected. They had been predicted by Einstein in 1916, and although indirect proof of their existence was found in 1974 by US physicists Russell Hulse and Joseph Taylor (they received the 1993 Nobel Prize for it), the waves themselves had remained elusive. Until now.

Thanks to a century-old prediction of Einstein, we have a new tool to look at outer space. A tool that responds not to light, but to something else: gravitational waves, tiny distortions of space and time that shoot through everything at the speed of light. Including the Earth. Including you. They make our time, and us, and everything, wiggle as they pass. Mankind has been blind to them since for ever. Not anymore.

But Einstein is not the only one. All the faces you saw in the stars, right after waking up on your beach, were the faces of the giants of the past and present. I obviously could not name them all, there are far too many, but these are the people whose legacy continues to make our world better known, and more vast, by the minute. They have built the story of our species. They have written, page after page, the book of what we know so far about our reality. Most of them are not known to the general public, but they are important nonetheless.

Remembering how your journey started, however, you may realize that you did not find a way to save the Earth from the future explosion of the Sun. You may not even have found a way to protect our planet from all the possible catastrophes that may happen before that. But you did discover what will enable our species to do so, and survive. Our brains. Our minds. Our imagination. Science.

As far as today's knowledge is concerned, it is impossible to travel from one part of the universe to another within a lifetime, or even a thousand. You could only do it in your mind. But only a few generations ago it took months to sail from Europe to Australia. It now only requires a few hours of flight. We do not know what tomorrow's technology will make out of theoretical work. We do not know what general relativity will one day allow us to do. Today, as I mentioned before, it has given us GPS. Just GPS. Tomorrow, it may allow us to find shortcuts in spacetime, the so called *wormholes* that could link two distant locations without having to cross the far-too-vast expanses that otherwise separate them. And you've seen that there are countless other planets out there, worlds that might one day welcome us . . .

So far, we humans have managed to travel beyond the clouds, to the Moon, and we have sent robots to the edge of the solar system. Beyond this edge, humanity has seen, rather than travelled, and you yourself surveyed everything that is known, and unknown, in a succession of gedanken experiments. Thanks to these mind trips, you have gathered together the sum knowledge of early twenty-first-century theoretical physics.

Some of what you have learned throughout this journey may turn out to be wrong, however. Dark matter, dark energy, cosmic inflation, parallel worlds or realities and strings are all ideas that may eventually be abandoned, but they are the most powerful ideas of our time nonetheless. They correspond to how mankind is trying, today, to make sense of this universe of ours. In a couple of centuries' time, all this may be discarded or accepted. We do not know. But to be alive today means being surrounded by these extraordinary ideas, for us to enjoy. So, before letting you think about all this on your own, here is a last summary of what you have seen, plus a bit more.

As you know, Newton did not figure out the ultimate theory of nature, the so-called and so far elusive theory of everything I alluded to some time ago, and for which string theory might be a contender. Newton's theory does not even explain Mercury's strange orbit, let alone the expansion of spacetime. So, in a sense, his theory is wrong. Still, it is quite brilliant. It can even be called perfect: we know where it works and we know where and why it breaks down. We are allowed to use it within (roughly) scales that can be grasped by our human brains: somewhere in between the very big and the very small, at velocities that are not too high, where the energies involved are not too intense. The world as we experience it, the world evolution has led us to detect through our senses, is contained within the valid limits of Newton's theory. Our common sense is rooted there.

But there are things that lie beyond. In the very fast, in the very small, in the very big or energetic. Within these beyonds, Newton's laws are of no use, and our senses are of no help, but still, amazingly, humankind has managed to unravel the laws of nature that apply where we cannot see. Quantum field theories apply within the very small, and general relativity takes over for the very large and energetically very dense. (And the very fast belongs to both.) In between these two, Newton is king. Where Newton doesn't work, strange new phenomena begin to be detected, and expected, hinting that new, mysterious realities border our own.

Both quantum field theories and general relativity have opened our eyes and minds to a universe far vaster than had ever been imagined by any of our ancestors, but still, these theories too have limits. Unlike with Newton's theory, however, no one knows for sure what lies beyond. Throughout this book, you have been journeying within these extraordinarily successful theories and, in the last part, you made a shy, tentative step beyond. You entered a universe whose basic constituents are made of strings and branes, a universe made of multiple realities and possibilities, of quantum vacua leading to strange laws in universes that are not our own.

Einstein's extraordinary vision was to see that gravity was not what Newton had thought it was. He showed that it was due to curves and slopes. Gravity and matter and energy are all linked in a very straightforward manner: our universe has a fabric, called spacetime, the curves and shapes of which are caused by what it contains, by what lies inside. The effect of these curves on nearby objects and light is what we call, what we experience as, gravity. That is the general theory of relativity. It is a hundred years old. To figure out the local shape of the universe outside a star, to figure out how its gravity affects its surroundings, one merely needs to know the energy contained within the star. Many scientists have done that calculation, starting with German physicist Karl Schwarzschild.

In 1915, the very year Einstein published his theory, at a time when only a handful of men and women around the world understood what it was about, Schwarzschild figured out the exact geometry of spacetime outside a star. Schwarzschild was forty-three years old at the time, and he achieved the feat while fighting on the Russian front during the Great War. He died a few months later of a disease he contracted there. Wars have deprived mankind of far too many individuals, including many who, like Schwarzschild, could have helped us understand the world better, and faster.

Following Schwarzschild's work, though, it was possible to guess how objects and light move around a star. It gave Mercury's correct orbit and showed that light itself should be deflected by the Sun. In 1919, an expedition led by British astronomer Sir Arthur Eddington detected such a (previously unnoticed) deflection. Photographs taken during a total solar eclipse that year showed that the stars near the Sun were not where they were supposed to be. Instead, they appeared precisely where Einstein's theory had predicted they would be, after being deflected by the Sun's effect on spacetime. Gravity being a curve in spacetime, light itself is subject to gravity.

Soon after Schwarzschild died, the same mechanism was applied to bigger objects still, galaxies, leading to the prediction that strange cosmic mirages, arcs of light floating in the middle of the faraway universe, exist. These were images of even more distant galaxies, whose light got distorted on its way to us. Galaxies, accordingly, were acting like cosmic lenses, allowing us to see behind them, to see further, deeper, into the history of our universe. Such lenses and mirages were detected more than sixty years after Einstein's work was published, in 1979. They can now be seen on almost every deep-space image taken by our telescopes. Incidentally they show that Einstein's geometrical interpretation of gravity works not only right next to the Sun, but throughout outer space.

General relativity has given us a new vision of the universe.

You, I, everybody and everything, we are surrounded by all the information that reaches us now, at this instant, from the past. We are sitting at the centre of our visible reality, and everything within this reality obeys Einstein's law, except within black holes. The same applies to our understanding of matter and light: the whole visible universe is ruled by the same laws that apply in our cosmic vicinity. The matter we are made of, the light that bounces off our skin, they all obey the same quantum laws, everywhere in our visible universe.

Linking faraway laws to nearby ones led to the discovery that our universe has a history, that it has a Big Bang in its past, that cosmic bygone eras can be read in the stars, using light, up to the point at which light cannot travel at all. This moment, this place in our universe's past when spacetime became large enough for light to travel freely, we named the surface of last scattering. The universe was 5,400°F hot when it vanished. Before that, the whole universe was opaque. After, it became transparent. What today remains of the temperature that radiated back then is what we call the cosmic microwave background. In it are imprints of what existed before.

Beyond that past, looking at the night sky has until now only led to indirect inferences about what once was. We may one day use our new eyes, the gravitational wave detectors, to receive signals from further away, but we are not there yet. Until then, we have to re-create the conditions that were once ubiquitous in the extremely tiny volume our universe was confined in during its infancy, in or-der to understand what happened.

Since the 1970s, particle accelerators have done just that. And they have led us to an unprecedented level of confidence in the the-ories used to probe the world of particles and light. Quantum field theories have given us a workable picture of what our universe is and was made of, up to a billionth of a billionth of a billionth of a second after the presumed birth of space and time as we know them, a birth whose existence is a prediction of Einstein's theory of general relativity.

And since the 1970s, we've also known that general relativity breaks down, that there are limits to what it can achieve. There, within its pitfalls, a new theory is needed, a theory of quantum gravity, and more. What that theory is, we do not yet know. (And there may even be many rather than just one.) But we do know that it some-how exists. That is what black-hole evaporation hints at.

As you shrank down to find where that new theory may lie, you ended up entering a new reality altogether, a reality made out of strings and branes and other dimensions. This was a step into string theory, perhaps the most popular contender for a theory of quan-tum gravity, or a theory of Everything, although it has yet to come up with predictions that could be experimentally checked.

It is within the landscape of these theories of strings and branes, sometimes referred to as *M-theory*, that the robot reached the end of its time as your guide through space, time and beyond, for you had entered a place where even the most powerful supercomputers invented by mankind cannot follow. Only human minds can reach it. There, you are free at last to figure out whatever you want about the world you live in.

There is hardly any doubt that discoveries to come, both theoret-ical and experimental, will reach further than today's knowledge, opening yet new windows onto a universe that is more extraordinary even than what any living being today imagines. General relativity and quantum field theories may then become perfect, like Newton's, for we shall know why they break down where they do, and what takes over. For now, however, they are wrong in the same sense that Newton was.

And thanks to these wrongs, we can peer into the unknown.

Without Newton, for lack of anything to compare it to, we wouldn't even have noticed Mercury's orbit's slight drift.

Without Mercury's disagreement with Newton's forecast and without Newton's inability to explain what happens when objects move very fast, we wouldn't have had Einstein's insight into how the universe's fabric interacts with its contents.

Without Einstein's equations, we would be like our ancestors, ignorant of the fact that our universe has a history. We wouldn't have built a picture of how our universe works as a whole.

Without this picture, you wouldn't have found gravitational waves and dark matter. Nor dark energy.

Wrongs are needed to find a right, to move forward.

Now, what about tomorrow? What will our new tool, the gravitational wave detector, change?

Four hundred years ago, when Italian physicist and philosopher Galileo pointed his newly invented telescope towards the heavens, he arguably became the father of observational astronomy.

And he saw that Jupiter had moons. He saw that there were celestial objects orbiting something that is *not* the Earth.

This shattered once and for all the millennia-old (mis)conception that everything revolved around our planet, that the Earth was at the center of the universe, paving the way for a scientific exploration of a reality unfathomably larger than expected.

Four hundred years later, Galileo's telescope has become the Hubble space telescope, X-ray telescopes, ultra-violet light and radio wave telescopes along with other light-based tools which have answered many questions about the cosmos, and our origins, eventually leading to the idea that our universe had not existed for ever.

But light does not propagate through everything. Just like you cannot see what lies behind a wall, we most of the time cannot see, using light, what lies on the other side of the Milky Way, or behind a far away galaxy, because dust and stars and, sometimes, other galaxies are in their path, putting us in their shadow. Not so with gravitational waves. They don't create shadows. Except behind black holes. A revolution of thought of similar proportion to Galileo's may hence be at hand: we have a new eye to watch the cosmos with.

The first gravitational wave ever recorded was the telltale signature of two black holes merging. We had no proof that black holes could orbit one another, let alone merge. That is already a discovery worth a Nobel Prize.

In the months and years to come, we will no doubt find many more black holes, maybe everywhere, of various sizes, and our theories about these strange cosmic monsters' lives will, at long last, be put to the test. From birth to death. Black hole interiors will still remain beyond experimental reach, though (once inside, even gravitational waves can't crawl out), but their surface, their horizon, can now be probed. Thanks to the signal caught in September 2015, it already seems that mankind was right about some of their properties, hinting that theoretical black holes do actually correspond to real ones: their size and shape depend only on very few parameters, namely their mass, their charge, and the way they spin on themselves. This is known as the black hole *no hair theorem*. It was stated (and called such) some 50 years ago by John Archibald Wheeler, the extraordinary physicist who supervised the PhD thesis of Richard Feynman, Hugh Everett III and . . . Kip Thorne, one of the founding fathers of the LIGO detector that caught these waves.

Thanks to the no hair theorem, black hole collisions and other spacetime storms will no doubt become ideal candles to estimate faraway distances, giving us an independent way to verify what has been inferred, up to now, using light only. The nature of dark matter and the existence of dark energy is in the balance here. We should soon know.

Now, if you are wondering about what is *not* expected, well, so do I! Will we see proof of extra-dimensions? Will we find something we've never thought of? Let's hope so. We've just built ourselves a new eye and the best a new eye can reveal is the unexpected, the unpredictable, to feed us with new mysteries to solve.

By 2017, three gravitational wave detectors should be up and running simultaneously: two in the US (that's LIGO) and one in Italy (that's VIRGO). For now, they can only detect gravitational waves whose sources are up to about 1.5 billion light years away. In a year's time, they should reach three times further. But there also is the LISA project, a European Space Agency space-based gravitational wave antenna way more powerful than LIGO and VIRGO. Its construction will no doubt get a nice boost now. My dream is that it will detect waves originating from beyond the surface of last scattering, through the opacity of our universe's tumultuous childhood. This would allow us to "see"—let's be optimistic—the inflation era (if it is real), the black holes that were born right after and, who knows, the Big Bang itself. Or, even better, something completely different, some wrongs to find a new right.

Next time you watch the stars and the Moon, I hope you will remember how strange and vast and beautiful this universe of ours is, for it is by widening our collective knowledge and dreams that, while hunting for hidden beauties and mysteries, we shall find the path to the long-term survival of our species.

Acknowledgments

Writing a book is no easy task. Less often considered, but equally true, is that writing a book is also a very selfish process.

For having allowed me to do it, and for helping me throughout this process, I am immensely grateful to Lauren, my beautiful shining stardust-made marvel.

Writing a book is one thing, but publishing it is something else again. I have many people to thank. In chronological order:

Philippa Donovan, at Smart Quill Editorial. Having read the proposal of my humble project (to write an "easy-to-read pop-science book about everything that is known about our universe from before the Big Bang up to today"), instead of quietly putting it into the bin, she introduced me to the best agent ever.

Antony Topping, at Greene & Heaton Literary Agency, is the best agent ever. And also the best friend a book, or an author, could hope for.

Jon Butler, who I hope knows, as I know, how much this book owes to him. His input has been creative, inspirational, gentle, incisive and, above all, understanding. I'm glad we still have a few unsettled theoretical issues to discuss – around a good many beers, I hope.

Kate Rizzo, at Greene & Heaton Literary Agency, thanks to whom this book is about to travel all around the world. And maybe even beyond. She is that good.

Everybody at Macmillan for their wit and enthusiasm. Without **Robin Harvie**, **Nicholas Blake** and **Will Atkins**, *The Universe in Your Hand* would never have been as readable as it is, and I never would have been as proud of it as I am.

Before I could give a copy of this book to my former supervisor, **Stephen Hawking**, I had to make sure there were no mistakes lingering in the text, and I am immensely proud to be able to thank my scientist friends who generously agreed to spend some of their precious time proofreading the science in the book: **David Tong**, professor of Theoretical Physics, Cambridge University, UK; **James Sparks**, professor of Mathematical Physics, Oxford University, UK; **Andrew Tolley**, assistant professor of Physics, Case Western Reserve University, USA; **Cristiano Germani**, Ramon y Cajal Researcher, Institute of Cosmos Science, University of Barcelona, Spain. I am indebted to you all.

Needless to say, I am the only one to blame if any mistake has managed to creep its way into the published book.

Having been able to give you, **Stephen Hawking**, a copy of the book, I will use this opportunity to express what an honour it is to be able to thank you: you introduced me to the wonders of theoretical physics. Everything I've learnt about our reality, I began learning from you: you taught me how to think about this beautiful world of ours, a world made even more beautiful by the existence of people like you.

Sources

For a book like *The Universe in Your Hand*, it is hard to describe exactly where the contents come from. I am not the one who discovered the theories, but I have done my best to interpret them.

I guess most of the material is rooted in graduate-student level textbooks and discussions I have had with Stephen Hawking and other dazzlingly brilliant professors.

Still, there is no doubt that what I know is also based on lectures and talks I attended while at the Department of Applied Mathematics and Theoretical Physics (DAMTP) at Cambridge University, UK, or while visiting the California Institute of Technology (Caltech), Pasadena, USA, or the Kavli Institute of Theoretical Physics, Santa Barbara, USA, where I used to spend about a month every year with Stephen and his other PhD students (Thomas Hertog, James Sparks and Oisín Mac Conamhna).

I cannot list all the scholarly articles that I've read on the arXiv while writing *The Universe in Your Hand*; they are far too numerous.

But here is a non-exhaustive list of some remarkable textbooks that I often browsed. Beware: these are not easy-to-read, popular science books. But they are great, and I'd like to record them here, as they have been so important to me.

Gravitation, by Charles W. Misner, Kip S. Thorne, John Archibald Wheeler (W. H. Freeman, 1973)

General Relativity, by Robert M. Wald (University of Chicago Press, 1984)

The Large Scale Structure of Space-Time, by Stephen W. Hawking and George R. Ellis (Cambridge University Press, 1975)

Black Hole Physics, by Valeri P. Frolov, Igor D. Novikov (Springer, 1998)

The Mathematical Theory of Black Holes, by Subrahmanian Chandrasekhar (Oxford University Press, 1998)

An Introduction to Quantum Field Theory, by Michael E. Peskin, Daniel V. Schroeder (Perseus Books, 1995)

Quantum Field Theory in a Nutshell, by A. Zee (Princeton University Press, 2010)

Quantum Fields in Curved Space, by N. D. Birrell and P. C. W. Davies (Cambridge University Press, 1984)

The Quantum Theory of Fields, vols. 1, 2 & 3, by Steven Weinberg (Cambridge University Press, 1995)

Superstring Theory, vols. 1 & 2, by Michael B. Green, John H. Schwarz, Edward Witten (Cambridge University Press, 1987)

String Theory, vols. 1 & 2, by Joseph Polchinsky (Cambridge University Press, 2000)

Quantum Gravity, by Carlos Rovelli (Cambridge University Press, 2007)

Euclidean Quantum Gravity, edited by Stephen W. Hawking, Gary W. Gibbons (World Scientific, 1993)

Index

51 Pegasi b, 41

absorption lines, 73
absorption spectrum, 73
"action at a distance," 153, 159,
    173
ageing, 144
air, 178–79
Al-Sufi, Abd Al-Rahman, 37
alpha particles, 198–99
Alpher, Ralph, 109
Anderson, Carl D., 231
Andromeda Galaxy, 37–39, 42, 43, 48,
    288
anisotropies, 336
anthropic principle, 354
anti-electrons (positrons), 230–31,
    266
antigravity force, 301, 332, 336
    see also dark energy
antimatter, 226, 227–33, 241, 290
anti-neutrinos, 231
antiparticles, 231–32, 247
anti-photons, 231
antipodes, 330–32, 334
anti-quarks, 231, 232, 267
anti-world, 232
Aspect, Alain, 216
asteroid belt, 22
asteroids, 14, 30–31, 86–89, 178, 237,
    308–12, 344
atmosphere
    Earth's, 93–94
    Moon's, 12–13
atomic clocks, 139–42
atomic energy, 200

atomic nucleus, 16, 19, 70–71, 163, 174,
    178, 181, 187–89, 196, 222, 223, 239
    electrical charge, 183
    splitting, 196, 198
atoms, 9, 15–19, 46, 70–74, 150–51,
    162–72, 173–82, 187–90, 195,
    198–99, 206, 340, 344
    electrical charge, 163
    forging of, 17, 177, 190, 192, 195, 206,
        223, 252
    and neutrinos, 197–98
    smallest, 238
    structure, 162–64, 170
    visualizing the surface of, 168–69, 173
axis, of the Earth, 36, 140

Balick, Bruce, 30n
ballistics, 248, 259
Bell, John Stewart, 216
Bell laboratories, 109
big, the very, 260–62, 263, 285, 305, 320,
    324, 336, 343, 345, 362
Big Bang theory, 47, 49, 99–100, 103–4,
    233, 234, 243, 246, 254, 261–62, 264,
    288, 295, 330, 332–34, 339, 364, 369
    cause, 49
    heat of, 99, 108–10
    multiple Big Bangs, 335–38
    time of the Big Bang, 112, 113
Big Crunch, 301
Binnig, Gerd, 168–69
black holes, 30–33, 34–36, 48, 95, 137, 205,
    237–38, 255, 261–62, 264, 286–87,
    296, 305, 341, 346, 349, 364–68
    black hole information paradox, 322
    death of, 318, 320

black holes *(continued)*
  ejection from, 32, 34
  gravitational effect, 55, 312,
    315–16
  gravity and, 55, 312, 315–16, 319
  heat, 319
  horizon, 309–18, 320–22
  inside, 307–26
  and mass, 320
  particle escape/evaporation from,
    317–23, 325, 341, 342, 349, 365
  Planck size, 325, 345–46
  radiation, 318, 320–23, 325–26
  shining of, 318–20
  shrinking of, 319
  small, 323–25, 345–46
  solar-mass, 323
  and string theory, 356
  supermassive, 29–30, 261–62, 286–87,
    323–24, 344
  temperature, 318–20, 323
  vacuum, 317
Bohr, Niels, 204
Born, Max, 204
branes, 355–56, 362, 365
Brown, Robert, 30n
Bruno, Giordano, 41, 96
bubble universes, 337, 356

cancer, 199
carbon, 16, 17, 178
Casimir, Hendrik, 218
Casimir effect, 219, 317
Cepheids, 294, 298
Chadwick, Sir James, 188
Charon, 23
CHNOPS, 178
Choquet-Bruhat, Yvonne, 95
cold death hypothesis, 301, 337
colour, 173

comets, 23, 24, 289, 344
  escape velocity, 90
  and water on Earth, 178
communications, 25–27, 41–42, 66–67
computers, quantum, 282
consciousness, 283
Copernicus, 39–40, 77
cosmic dance, 37–39
cosmic Dark Ages, 46, 48, 78, 238,
  292
cosmic horizon, 102
cosmic microwave background
  radiation, 110, 236–28, 245–46, 323,
    329–32, 335–36, 364
cosmogonies, 98
cosmological constant, 300–301
cosmological inflation, 244
cosmological principles, 99, 103
  first, 63, 73
  second, 78
  third, 78–79
cosmology, 96–100
  observational, 80
Curie, Marie, 194
currents, 158

Dark Ages of the universe, 46, 48, 78,
  238, 292
dark energy, 292–302, 322, 344, 361,
  366, 368
dark matter, 286–91, 292, 299, 302, 322,
  344, 361, 366, 368
deoxyribonucleic acid (DNA), 178,
  199
diamond, 310–12, 320
dinosaurs, 10, 288
  extinction of, 237
Dirac, Paul, 228–31, 234, 343, 357
distance, 234
  estimates, 293–94

observer-dependent nature, 134–35,
141–42
and speed, 125, 128–30, 132–36, 141–42
double-star systems, 40, 41, 120
dwarf planets, 23

E=mc², 18, 92, 94, 132, 190, 224, 229,
232, 335
Earth, 21, 26–27, 329
atmosphere, 93–94
axis, 36, 140
as centre of the visible universe, 49, 367
core, 15, 16–17, 28
cosmic horizon of, 102
and the creation of the Moon, 12–14,
237
escape velocity, 89
exploration, 227–28
"flat Earth" hypothesis, 39
galactic position, 288
gravity, 16–17, 55, 82–83, 89–94, 141,
184, 206, 259
layers, 90
life on, 14, 17, 178, 200
movement through space, 122
obliteration by the Sun, 6–8, 9
orbit of the Sun, 39, 54, 55, 56–57, 293
past, 237
position in the Milky Way, 35, 36
position in the universe, 77
and radioactivity, 200
tides, 13, 94
"Earthlike" planets, 40–42
Eddington, Sir Arthur, 363
Einstein, Albert, 71, 76, 81, 93, 94,
96–99, 103, 107, 108–10, 112, 204,
228, 261, 263, 288, 304, 324, 352, 366
on dark energy, 300
E=mc², 18, 92, 94, 132, 190, 224, 229,
232, 335

gedanken experiments, 359
general theory of relativity, *see*
general theory of relativity
on gravity, 60, 125, 153
gravitational waves predicted by, 95,
360
as patent officer, 125
on the quantum world, 168
special theory of relativity, 126–27,
132–33, 135, 139, 140, 142, 145, 359
and time, 125–27
electric charge, 183, 230, 265
electric currents, 169
electricity conduction, 195
electromagnetic charge, 160–61, 174,
180
electromagnetic field, 157–61, 163,
164–65, 167, 170, 173–74, 180, 187,
191, 220–21, 231, 239, 269
and electrons, 161, 167, 170–72, 174,
180, 220, 230, 265
fundamental particles of, *see*
electrons; photons
negative charges, 160
positive charges, 160
vacuum of, 229
electromagnetic force, 159, 169–70, 184,
191, 350
electromagnetic force carrier, 380
electromagnetic radiation, 64
see also light
electromagnetism, 184, 205, 351
electron microscopes, 169
electrons, 16, 19, 70, 163–67, 173–82,
187, 191, 199, 206, 210, 224, 229,
239, 265–67, 269–70, 275, 344
appearance, 167–69
broken bonds, 179
electrical charge, 229–30, 265
and electricity conduction, 195

electrons *(continued)*
  and the electromagnetic field, 161, 167,
    170–72, 174, 180, 220, 230, 265
  as fundamental particles, 170
  as identical to each other, 164, 167,
    174, 177, 219–20, 265
  loss of, 179
  movement, 164–70
  orbitals, 70–72, 175, 180
  as particles, 175
  and the Pauli exclusion principle,
    175–76, 179, 180–82
  and photons, 170–72
  position, 167
  velocity/speed, 141–42, 167
  very energetic, 198–99
  as waves, 175
electroweak field, 205, 222, 225, 242
elliptical orbits, 56
emission spectrum, 72–73
emptiness, lack of, 217–19
energy
  at the beginning of the universe, 240,
    242–44
  beginning to turn into particles, 243
  dark, 292–302, 322, 344, 361
  E=mc², 18, 92, 94, 132, 190, 224, 229,
    232, 335
  and electron orbitals, 71–72
  infinite, 267
  kinetic, 87
  and mass/matter, 92, 93, 107, 133,
    136, 137, 172, 190, 229–30, 232,
    242
  Planck, 346
  quanta, 269–70
  release through breaking electron
    bonds, 179–80
  and size, 346–47
  of the Sun, 15–20, 21, 63

too much in a small volume, 310–11
  of the visible universe, 99
Englert, François, 225–26
escape velocity, 89
eternal inflation, 337, 339, 343, 354
Euclid, 293–94
European Centre for Nuclear Research
  (CERN), 182, 225, 241
European Space Agency, 23, 89n
Everett III, Hugh, 283–85, 337, 368
evolution, 117
exoplanets, 41–42
extinction, 10, 65, 121
extraterrestrial life, 40–42
eye, 65, 101, 117

fast, very, world of, 119, 123–24, 125–38,
  139–45, 260
fastest known object in the universe,
  28–29
Feynman, Richard, 204, 272, 317, 343,
  368
fields, 153, 234, 242, 243–44, 250, 252, 265
  see also quantum fields
Fields Medal, 207n
force carriers, 159, 170, 184, 187–89, 191,
  193, 196–97, 222–23, 224, 240–41,
  250
  of the electroweak field, 242
  gravitational, 220, 270–71, 351
forces, 153, 155–61
  anti-gravitational, 301, 332, 336
  electromagnetic, 159, 169–70, 184,
    191, 350
  strong nuclear force, 186, 189, 196,
    241, 270
  unification, 205
  weak nuclear force, 193, 197, 205
"fourth force," *see* gravity
freezing death hypothesis, 301, 337

fundamental particles, 170, 184, 219–20,
    222, 225, 240, 246, 269–70, 271, 315,
    318, 321–22
  gravitons, 220, 270–71, 351
  of the inflaton field (inflatons),
    332–33, 335
  neutrinos, 197–98, 220, 224
  see also electrons; gluons; photons;
    quarks
future, 131
  time travel to, 123, 142–45, 208,
    250–55, 312–13, 359

galactic year, 288
galaxies, 27, 34–42, 43–46, 97, 101,
    237–38, 286, 344
  centre, 28, 35, 48
  clusters, 43–44, 97
  and dark energy, 292
  and dark matter, 290–91, 292
  distance estimates, 294
  and gravity, 80, 94, 363
  growing distances between, 75–78,
    80–81, 104
  superclusters, 43, 44
  see also Andromeda Galaxy; Local
    Group; Milky Way
Galileo Galilei, 39–40, 366–67
gamma rays, 32, 65, 72, 199
Gamow, George, 109
gases, 179
gedanken experiments, 359, 361
general theory of relativity (Einstein's
    theory of gravity), 60–61, 93,
    118–19, 141, 142, 145, 203, 246, 254,
    271–72, 286–87, 291, 298, 299–300,
    301, 315, 342, 360, 362–66
  breakdown of, 264, 304–6, 314,
    315–16, 365–66
  gravitational waves and, 95

genetic mutations, 199
Glashow, Sheldon Lee, 205, 222–24
Global Positioning Systems (GPS), 142,
    361
gluons, 185–91, 193, 196, 219, 220, 222,
    224–26, 231, 242, 243, 267, 270, 344,
    356
gods, 10, 203
gold, 19, 149–51, 155, 162, 169, 174, 176,
    187, 191, 194–96
"Goldilocks Zone," 26
grand unified field, 243
Grand Unified Theory, 222
grand-unification scale, 346, 349
gravitational constants, 324
gravitational field, 220–21, 223, 270
  basic quanta, 269–70
gravitational-wave detectors, 69, 245,
    360–68
gravitational waves, 94–95, 254, 360–367
gravitons (force carriers), 220, 270–71, 351
gravity, 9, 11, 81, 82–95, 120, 135, 153,
    184, 206, 220–21, 223, 250, 267, 286,
    319, 344
  at the beginning of the universe,
    240–45, 249
  black holes and, 55, 312, 315–16, 319
  as curvature/bending of spacetime,
    58–61, 88–94, 107, 118, 135–36, 153,
    241, 244–45, 286, 290, 303, 312, 362
  definition, 57, 59–60
  Einstein's theory of, *see* general
    theory of relativity
  and the forging of atoms, 190
  galaxies and, 80, 94
  and inflation, 336
  and neutrinos, 197
  Newton's universal law of
    gravitation, 54–57, 59, 118, 259, 287,
    289, 291, 298, 301–2, 305, 324

gravity *(continued)*
  not a force, 90, 92, 118, 153
  and the Planck scale, 347
  and quantum theory, 244–46, 254,
    259–64, 267–68, 269–72, 305–6,
    308, 316, 320, 322, 324–25, 339,
    340–41, 344–47, 350, 365
  and string theory, 350–51, 355
  of the Sun, 17, 19–20, 21, 54, 55, 94, 99
  and time, 141, 144
Gross, David, 186
Guth, Alan, 332

hadrons, 225n
Hafele, Joseph, 139–41
Haroche, Serge, 282–83
Hartle, James, 341–42
Hartle–Hawking wave-function of the
    universe, 342
Haumea, 23
Hawking, Stephen, 40n, 60, 204, 205,
    251, 272, 304–6, 315, 319, 321–23,
    325, 339, 341–43, 345
Hawking radiation, 320, 323
Hawking temperature, 320, 323
heat, 99, 109–10, 221, 238–39
Heisenberg, Walter, 277
Heisenberg uncertainty principle, 277–78
helium, 16–17, 46, 73, 82, 175, 198, 238,
    239
Herman, Robert, 109
Higgs, Peter, 225–26
Higgs field (Higgs–Englert–Brout field),
    225–26, 242, 344
Higgs particle, 225, 242, 254
horizons
  black hole, 309–18, 320–22
  cosmic, 102
Hoyle, Fred, 110
Hubble, Edwin, 37, 40, 80, 294, 300

Hubble's law, 80, 81, 294–95, 298
Hulse, Russell, 360
human beings, 44, 117
hydrogen, 16, 17, 20, 46, 73, 162, 163,
    165, 167, 169–71, 174–78, 183, 187,
    191, 192, 195, 238, 239, 275

IBM, 168
infinities (singularities), 304–5, 314, 319,
    356
infinities, quantum, 214, 265–73, 274,
    276, 278–85, 303, 317, 335, 345, 351
  renormalization (removal), 269, 270,
    345
inflation, eternal, 337, 339, 343, 354
inflation era/epoch, 332–34, 335, 339,
    343, 345, 354, 356
inflaton field, 243–44, 332–34, 335–36, 344
inflatons (fundamental particles of the
    inflaton field), 332–33, 335
infrared light, 65, 75, 111
interconnectedness, 253
International Space Station, 130
ions, 19, 179

Jupiter, 12, 22, 39, 40, 41, 55, 208, 209, 366

Kant, Immanuel, 38n
Keating, Richard, 139–41
Kepler 186f, 41
Kepler telescope, 41
kinetic energy, 87
Kuiper belt, 22

Lamoreaux, Steve, 219
Langevin, Paul, 126–27, 129
Large Hadron Collider (LHC), 225, 241,
    347n
lasers, 181
Lemaître, Georges, 99

length, 324
  contraction, 129–30, 132–34
  observer-dependent nature, 134,
    141–42
  Planck, 262, 324, 340, 347
  and speed, 125, 129–30, 132–35, 141–42
LHC (Large Hadron Collider), 225, 241,
  347n
life on Earth, 14, 17, 178, 200
light, 63–69, 180–81, 187, 198, 210–11,
  238, 329, 349
  absorption spectrum, 73
  at the beginning of the universe, 241
  and black holes, 308, 311, 314, 319,
    346
  and colour, 173
  deflection, 88, 363
  and the electromagnetic field, 160–61
  and electrons, 71–72
  first, 46, 68
  as frozen record of time/the past,
    66–69, 70, 106, 143, 251–52, 329
  furthest detectable, 45, 68
  and gravity, 91
  inability to travel through dense
    spacetime, 107, 110, 239
  intensity (height of a wave), 64
  light pollution, 12n, 193
  and matter, 171, 173, 181
  and our knowledge of the universe,
    66–70, 364
  packet theory, 71n
  and the Planck scale, 347
  redshift, 74, 75, 143n
  star light, 192–93
  and string theory, 350
  and the surface of last scattering, 47,
    48–49, 79, 101, 108, 111, 113, 364
  travel through space, 88
  ultraviolet, 65, 367

  virtual pearls of (virtual photons),
    159–60, 164–65, 169n, 181, 183–84,
    186, 191, 196, 265–66
  visible, 65
  as a wave, 275
  wave frequency, 64
  wave–particle duality, 64
  wavelength, 64–65, 72, 75, 111, 276
  see also photons
light speed, 18, 27, 67–68, 134, 141–43,
  224, 324, 332
  $E=mc^2$, 18, 92, 94, 132, 190, 224, 229,
    232, 335
  expansion of the universe at a speed
    faster than, 332
  and mass, 143
  moving at percentages of, 128–32,
    133–34, 140–41, 142
  speeds greater than, 62, 332
LIGO, 95, 368
Linde, Andrei, 332
liquids, 179
LISA project, 368
Local Group (galaxy group), 43, 48, 94
longevity, 145

M-theory, 365
magnets, 57, 152–55, 157, 159, 160, 161,
  170, 176, 181, 184, 191, 222, 265
Makemake, 23
Maldacena, Juan, 355, 357
Mars, 12, 13, 21, 55
mass, 18, 190, 223–26, 234, 324, 344
  and black holes, 320
  $E=mc^2$, 18, 92, 94, 132, 190, 224, 229,
    232, 335
  and energy, 92, 93, 107, 133, 136, 137,
    172, 190, 229–30, 232, 242
  Planck, 324, 325
  and quarks and gluons, 186

mass *(continued)*
  and speed, 131–33, 135–37, 143
  and time, 143
masslessness, 143
mathematics, 203, 204
  language of, 53–54
Mather, John C., 336
matter, 211, 226, 241, 299, 349, 364
  antimatter, 226, 227–33, 241, 290
  composition, 15
  dark, 286–91, 292, 299, 302, 322, 344,
    361–68
  and energy, 92, 93, 107, 133, 136, 137,
    172, 190, 229–30, 232, 242
  humanity's understanding of, 182
  and light, 171, 173, 181
  Milky Way's shortage of, 288–89
  and the Planck scale, 347
  and spacetime, 303
  and string theory, 351
  too much in a small volume, 304–6,
    310–11, 314
Mayor, Michel, 41
Mercury, 21
  orbit of the Sun, 55–57, 58, 59, 287,
    289, 305, 361–63, 366
mesons, 189, 190, 193
metals, 71
microbes, 344
microscopes
  electron, 169n
  scanning tunnelling, 169
microwaves, 65, 72, 111, 117
  cosmic microwave background
    radiation, 110, 236–28, 245–46, 323,
    329–32, 335–36, 364
Milky Way, 11, 12, 27, 32, 33, 34–42, 43, 44,
    45, 48, 80, 97, 101, 193, 286–91, 295
  bending of, 94
  black hole of, 307

collision with Andromeda, 39, 288
  cosmic dance, 38–39
  distance estimates, 293–94
  gravitational field, 94
molecules, 92, 177–80, 191, 194, 199, 344
Moon, 94, 361
  atmosphere, 12–13
  craters, 12–13
  creation, 12–14, 237
  dark side, 12
  and Earth tides, 13, 94
  gravity, 94
  orbit of the Earth, 54
multiverses, 104, 260, 337
  beginnings of, 341–42
mystics, 234

NASA, 23n, 25n, 41, 299
Nature, 319
nature, laws of, 63, 73, 103, 212–13, 214,
    226, 243, 285, 342, 354, 359
nature's constants, *see* gravitational
    constant; Planck's constant; speed
    of light
nebulae, 177
neon, 175
Neptune, 22, 55
neutrinos, 197–98, 220, 224
neutron stars, 296
neutrons, 188–91, 295–96, 198, 224, 226,
    240–41, 291, 344
New Horizon spacecraft, 23n
Newton, Isaac, 10–11, 53–57, 59, 82, 118,
    134, 153, 229, 248, 259, 352
  and God, 11
  and gravity, 54–57, 59, 118, 259, 287,
    289, 291, 298, 301–2, 305, 324,
    361–66
  and a theory of Everything, 361
nitrogen, 16, 178

no-boundary proposal, 341
Nobel Prize, 80, 181, 207n, 226n
Nobel Prize in Chemistry, 183, 194,
      197
Nobel Prize in Physics, 169, 203, 272
      1903, 194
      1918, 324
      1921, 71, 204
      1922, 204
      1932, 277
      1933, 231, 279
      1935, 188
      1936, 231
      1945, 176
      1949, 189
      1954, 204
      1965, 204
      1978, 110
      1979, 205
      1982, 269
      1986, 169
      1993, 360
      1999, 268
      2004, 186
      2006, 336
      2011, 299, 334
      2012, 282
      2013, 225
nothingness, 183, 314
nuclear power plants, 200
nuclear weapons, 200, 232

observational cosmology, 80
ocean floor, 228
Oort, Jan, 288–91
Oort cloud, 24, 25, 289
Öpik, Ernst, 37, 40, 80
osmium, 296
outer space, 155
oxygen, 16, 17, 19, 176–78, 195

parallel processing, 282
parallel universes, 9, 263, 284–85,
      355–56, 361
particle accelerators, 136–37, 222–23,
      225, 245, 325, 364
particle–antiparticle pair annihilation,
      230–33, 266, 317
particle–antiparticle pair creation,
      230–31, 266, 317
particles, 180, 209–21, 228, 246, 250,
      252, 263, 265, 267–68, 272, 274,
      276–79, 280, 283, 303, 316, 352
   and Heisenberg's uncertainty
      principle, 277
   Higgs particle, 225, 242
   mass, 224–26, 229
   massless, 224
   particle–wave duality, 219
   and string theory, 351
   superposition, 283
   virtual, 217–18
   see also specific particles
past, 61, 62, 79, 98, 251, 340, 364
   and the Big Bang, 104
   and the expansion of the universe,
      81, 94–95, 104, 106, 110
   looking into, 106–10, 112–13, 143
   and the surface of the last scattering,
      101
   time travel to, 235–46, 251–52, 260, 262
Pauli, Wolfgang, 176, 179, 197
Pauli exclusion principle, 175–76, 179,
      180–82
Penrose, Roger, 304–6, 315
Penzias, Arno, 109
Perlmutter, Saul, 298, 299, 301, 334
PET (Positron Emission Tomography),
      232–33
Philae space probe, 23, 89
phosphorus, 178

photoelectric effect, 71n
photons, 180, 219, 220, 239, 269–70, 344
    and electrons, 170–72
    scattering, 110
    very energetic, 199
    virtual, 159–60, 164–65, 169n, 181,
        183–84, 186, 191, 196, 265–66
Planck, Max, 324–25
Planck energy, 346
Planck length, 262, 324, 340, 347
Planck mass, 324, 325
Planck scale, 262, 341, 345–47, 349
Planck time, 325
Planck wall/era, 245, 255, 259, 263, 264,
        324, 339–41, 343, 345
Planck's constant, 324
planets, 21
    collisions, 13–14
    dwarf, 23
    "Earthlike," 40–42
    exoplanets, 41
    gas giants, 21, 22
    obliteration, 6–8, 9
    observing those outside our solar
        system, 41–42
    orbits, 39–40, 54, 55–57, 58, 59, 70,
        287, 289, 293, 305, 361–63, 366
    rocky, 21
    slingshotting off, 87–90
    see also specific planets
plasma, 6, 15, 18, 19, 63
Plato, 203
Pluto, 23
plutonium, 151, 195–96, 198
plutonium-239, 195
Politzer, David, 186
polonium, 194, 198
position, uncertainty of, 277
Positron Emission Tomography (PET),
        232–33

positrons (anti-electrons), 230–31, 266
possibilities, 216, 266, 276, 278–85, 317
    superposition of, 281–83, 340
protons, 185, 187, 189, 190, 191, 195, 196,
        198, 224–26, 241, 291, 344
Proxima Centauri, 24, 25–27, 34, 40, 289
    core, 26

quanta, 269–70
quantum computers, 282
quantum field theory, 254, 267–72, 285,
        303, 361–66
quantum fields, 153, 161, 166–69, 184,
        191, 193, 205, 206, 210, 215, 217–22,
        225, 231, 240, 246, 249, 263, 265,
        267–68, 270–71, 274, 318, 332, 337,
        343, 344–45, 346, 349, 351
    creation of the idea of, 228
    electroweak field, 205, 222, 225, 242
    inflaton field, 243–44, 332–34, 335–36,
        344
    and quantum jumps/tunnelling,
        166
    strong interaction quantum field, 185,
        186–88, 191, 220, 222, 225–26
    unified, 222–23, 233, 254
    weak nuclear quantum field, 193, 197,
        220
    see also electromagnetic field
quantum jumps/tunnelling, 165–66,
        316–17, 322
quantum wave collapse, 276, 278, 282,
        283
quantum world, 149–200, 203, 207,
        210–22, 260–64, 265–73, 274–85,
        323–24, 340–41, 342, 352
    see also small, very, world of
quark–antiquark pair, 267
quark jails, 185, 187–90, 195–96, 198,
        224, 241, 242

quarks, 184–91, 197, 219n, 222, 225, 226, 231, 242, 267, 270, 344
   up and down, 188, 196
Queloz, Didier, 41

radiation
   black hole/Hawking, 318, 320–23, 325–26
   cosmic microwave background, 110, 236–38, 245–46, 323, 329–32, 335–36, 364
   electromagnetic, 64
radio waves, 65, 111
radioactive burns, 199
radioactive decay, 197–99, 206, 220, 232, 280
radioactivity, 193–94, 197, 198–200, 222
radium, 194
reality
   beginning of, 98–100
   control of, 216
   human perception of, 118
   observation-dependent nature of, 214
   unknowability of, 96
   visible, 113
red dwarves, 25, 26, 40
red giants, 296–97
redshift, 74, 75, 143n
Rees, Sir Martin (Baron Rees of Ludlow), 58
Reines, Frederick, 197
Reitze, David, 95
relativity, *see* general theory of relativity; special theory of relativity
renormalization (removal of infinities), 269, 270, 345
Riess, Adam, 298, 299, 301, 334
Rohrer, Heinrich, 168–69
Rome, 41
Rosetta spacecraft, 23

Ruska, Ernst, 169n
Rutherford, Ernest, 183, 188

S2 (Source 2), 28–29, 30, 31, 288
Sagan, Carl, 40
Sagittarius A*, 30, 287
Salam, Abdus, 205, 222–24
satellites, 22, 41, 90, 109, 142, 178, 227, 299
Saturn, 22, 55
scanning tunnelling microscope, 169
Schmidt, Brian, 298–99, 301, 334
Schrödinger, Erwin, 359
Schrodinger's cat, 279–82, 284, 335
Schwarzschild, Karl, 362–63
senses, 117–18, 173
silver, 16
singularities, 344–7, 356–7
small, very, world of, 119, 145, 149–200, 203, 212, 214, 216, 219, 259–62, 263, 265–67, 268, 274–75, 277, 283, 285, 286, 301, 305, 320, 324, 336, 343, 346–47, 361
   see also quantum world
Smoot, George F., 336
solar system, 24, 25, 34, 289
   centre of, 28
   position in the Milky Way, 35–36
solar winds, 192–93
solidity, 176
space, 250, 260
   beginning of, 234
   bending of, 58–61, 88–95
   origin of, 203–55, 285
   and the Planck wall/era, 339
   and speed, 125, 128–30, 132–36
   see also spacetime
space travel, 62, 193, 360–61
spacetime, 113, 135, 141–42, 206, 250, 271, 306, 352, 362–368
   age of, 245

spacetime *(continued)*
   bending of, 93, 117–18, 135–36, 141,
      153, 264, 304
   beyond the surface of last scattering,
      112, 364
   and the Big Bang theory, 99–100
   birth of, 112
   and black holes, 260–61
   breaking of, 304
   changing/evolving nature of, 300
   distortion in a black hole, 308–15, 318
   expansion at a speed faster than light,
      332, 335
   as the fabric of the universe, 93
   fundamental packets of, 271–72
   growth of, 107, 109, 111
   light's movement through, 101
   and the surface of last scattering, 101
   total absence of, 99
   travelling beyond the origins of, 255
   and wormholes, 361
special theory of relativity, 126–27,
      132–33, 135, 139, 140, 142, 145, 359
spectrum, 73
speed
   and mass, 131–33, 135–37, 143
   as non observer-dependent, 134
   relativistic, 139
   and space, 125, 128–30, 132–35
   and time, 124, 125–33, 135–36,
      139–44
   and time travel, 122–23, 131, 142–45
   see also light speed; velocity
Starobinsky, Alexei, 332
stars, 11, 27, 28, 34–36, 45, 48, 192,
      241–42, 344, 349
   atom formation, 17, 192, 194–95, 206,
      223, 238
   Cepheids, 294, 298
   collisions, 39

cores, 17–18
dead, 67–68
death of, 5–7, 9, 17, 30–32, 177, 192,
      194–95, 237–38, 305, 309–10,322
double-star systems, 40, 41, 120
energy, 17–18
first-generation, 46
forging of, 177, 190, 252
and the geometry of external
      spacetime, 363
gravity, 94, 193
neutron, 296
past light of, 106
position of, 286
red dwarves, 25, 26, 40
red giants, 296–97
second-generation, 17, 46
shining of, 19, 189, 224, 310
size and lifespan of, 26
space travel to, 335–36
supernovae, 45, 298
third-generation, 17, 46
velocity of, 286–90, 294, 312
visible to the human eye, 11
white dwarves, 296–97, 311, 349
see also Sun
stellar graveyards, 28
stellar nurseries, 28
string theory, 207n, 262, 337–38, 349–58,
      361, 362, 365
strong interaction quantum field, 185,
      186–88, 191, 220, 222, 225–26
strong nuclear force, 186, 189, 196, 241,
      270
   fundamental particles of, *see* gluons;
      quarks
sulphur, 178
Sun, 14, 15–20, 21–24, 27, 28, 63, 287
   as a black hole, 323
   core of, 15, 16, 18–19

death of, 5–7, 9, 15, 19, 39, 288
Earth's orbit of, 39, 54, 55, 56–57, 293
energy of, 15–20, 21, 63
escape velocity, 89
gravity of, 17, 19–20, 21, 54, 55, 94, 99
light of, 68
location within the Milky Way, 35
as second- or third-generation star, 46
shining of, 19
size of, 15
temperatures, 16
supernovae, 45
type Ia, 298
surface of last scattering, 47, 48–49, 79,
    101, 108, 111, 113, 236–39, 245–46,
    329, 364
heat of, 110
and the Planck wall, 339

't Hooft, Gerard, 268, 270, 357
Taylor, Joseph, 360
telescopes, 73, 106, 366–67
ultraviolet light, 29
theoretical physics, 53, 203, 207, 221,
    245, 272, 355, 359, 361
theory of Everything, 254–55, 259, 262,
    264, 269, 302, 351, 361, 365
thermonuclear-fusion reactions, 18–19,
    162, 189, 192, 238
Thorne, Kip, 95, 368
thought (*gedanken*) experiments, 359, 361
tides, 13, 94
time, 234, 250, 260, 324
bending of, 93
freezing of, 143
imaginary, 341–42
and mass, 143
observer-dependent nature of, 134,
    141, 144
origin of, 203–55, 285

Planck time, 325
and the Planck wall/era, 339
and speed, 124, 125–33, 135–36, 139–44
universal, 125, 133, 140, 142
time-dilation, 126–27, 133, 139–41
time travel, 122–23, 131, 142–45
to the future, 123, 142–45, 208,
    250–55, 312–13, 359
to the past, 235–46, 251–52, 260, 262
time zero, 339

ultraviolet light, 65
uncertainty, 216
universe, 62, 259–64
age of, 61, 239, 246
beginning of, 99–100, 113, 236–46,
    260–62, 342
bubble universes, 337, 356
changing nature of, 61, 75–78, 80–81
edge of the visible universe, 47, 101,
    110–11, 240, 297, 329
end of, 45
expansion of, 75–78, 80–81, 98, 101,
    103–5, 106–11, 113, 143, 236,
    237–38, 244, 294, 298, 300–301,
    329–34, 335, 344
fabric of, 58–60, 75–78, 80–81, 87–88,
    91–93, 96, 113, 135
infinity of, 104
multiverses, *see* multiverses
mysteries of, 9
opaque, 46, 108, 110, 113, 263, 364
parallel, 9, 263, 284–85, 355–56, 361
shape of, 60, 92, 96
size of, 9–10, 40, 43–47, 99, 101, 103–5,
    106–7, 232, 237, 241–42, 244, 245,
    343, 344
and string theory, 352–57, 362
temperature of, 99, 109–11, 238–43,
    329–31, 334, 335, 364

universe *(continued)*
    ten-dimensional, 353
    transparent, 46–47, 68, 101, 109,
        111–12, 238, 240, 329
    violence of, 13, 292
    visible, 38, 40, 43, 46, 49, 68–69, 77,
        81, 97–98, 99, 101–4, 106, 109–13,
        237–40, 297, 329, 343, 364
    without beginning, 342
    without size, 99
    unknown, 263, 269
    uranium, 198
    Uranus, 22, 55

vacuum, 207–8, 218–20, 229, 233, 265,
        267, 271, 303, 349, 362
    black hole, 317
    different types of, 219, 221
    force of, 219
    of the inflation field, 332–33, 336
    quantum fluctuations, 267
    velocity, 90
    of electrons, 141–42, 167
    escape velocity, 89
    of stars, 286–90, 294, 312
    uncertainty of, 277
    see also light speed; speed
Veltman, Martinus, 268, 270
Venus, 12, 21, 41, 55
VIRGO, 368
vision, 173
    infrared, 65
    ultraviolet, 65

voids, 196, 234
Voyager 1, 25n

W bosons, 197, 223, 242
water, 178, 183
    on Earth, 178
    liquid, 26
wavelength, 64–65, 72, 75, 111, 276
weak nuclear force, 193, 197, 205
weak nuclear quantum field, 193, 197,
    220
Weinberg, Steven, 205, 222–24,
    357
Weiss, Rainer, 95
Wheeler, John Archibald, 283, 368
white dwarves, 296–97, 311, 349
Wilczek, Frank, 186
Wilkinson Microwave Anisotropy
    Probe (WMAP), 299n
Wilson, Kenneth Geddes, 268–70
Wilson, Robert, 109–11
winds, solar, 217
Wineland, David J., 282–83
Witten, Edward, 207, 215–16
wormholes, 361
Wright, Thomas, 37n–38n

X-rays, 32, 65, 367

Yukawa, Hideki, 189

Z bosons, 197, 223, 242
Zwicky, Fritz, 290